Lessons
from the
School
of
Hardknocks

Lessons
from the
School
of
Hardknocks

Brother Brooklyn

XULON ELITE

Xulon Press
2301 Lucien Way #415
Maitland, FL 32751
407.339.4217
www.xulonpress.com

Xulon Elite

© 2022 by Brother Brooklyn

Contribution by: BKX Studios Brandon Coley, Kenya Coley
Dr Jaiden Cole, Adriene Cole Esq
Ebony Smith, Denise Watts

All rights reserved solely by the author. The author guarantees all contents are original and do not infringe upon the legal rights of any other person or work. No part of this book may be reproduced in any form without the permission of the author.

Due to the changing nature of the Internet, if there are any web addresses, links, or URLs included in this manuscript, these may have been altered and may no longer be accessible. The views and opinions shared in this book belong solely to the author and do not necessarily reflect those of the publisher. The publisher therefore disclaims responsibility for the views or opinions expressed within the work.

Paperback ISBN-13: 978-1-66284-541-3
Ebook ISBN-13: 978-1-66284-542-0

Dedicated to my Mother, Since my conception.
To my children, for being able to See for themselves.
To my God Brother MKW,
"Don't be like me, be better than me."
To the human beings on lock down because
they did not know better.
To the awareness of mental manipulation.

Preface

I AM KNOWN as Brother Brooklyn, as I was born in Brooklyn, New York, in 1960. I quickly developed a "Brooklyn Hustle," mastering making money in legitimate and illegitimate ways. Hustling for money was how I saw members of my community getting what they wanted in life. Eventually, I also adopted this lifestyle to get the things that I wanted, things I thought pleased me, and things that made me happy.

Money is one consistent thing that people will go to great lengths to get. As I think back, the original reason I wanted money was to make my mother happy. My household always seemed happier when we had money, and naturally, I developed neural pathways that correlated happiness with money.

There was a period in my life when the only consideration for making decisions was whether or not I could make money. During that time, I always had to ensure that my actions would manifest a specific external reward. My thought process was, "If I do this, then this is the reward I should receive." Once I formulated the reward associated with a particular effort, I manifested my thoughts into action.

In the early 1990s, crack cocaine was starting to get popular. I figured out that I could purchase guns in Virginia, bring the guns up to New York to sell for a nice profit, then use that profit to buy cocaine in New York, and take that cocaine back down south, cook it up, sell it, and make more profit. This was a vicious cycle. In the beginning, it was all profit. Then, I slowly started cutting into my profits as I began using the cocaine that I was supposed to sell.

My Inner man's conscience informed me about rule number four "never get high off your own supply" and that this was the wrong thing to do. But the trap had been "sprung" and in my God given creativity, I created a master plan to get more money than I had ever had and maintain that subconscious drug habit drug habit that I never had as well.

The plan was to understand what things could get me caught and be prepared for those things. For example, As I traveled back and forth on the New Jersey turnpike, and I knew that, at the time, the police did a lot of profiling, so I made sure I did not fit that profile. I maintained my U.S. Navy look with a neat and trim haircut: looking the part that I used to be a part of. This misguided logic kept me from ever getting detained and kept me out of trouble for a while.

The diversified opportunities of living in New York City provided my "Brooklyn Hustle creativity opportunities to continue the simple math of one dollar plus one dollar equals two dollars. I was able to take advantage of the "white collar crime" that I saw other humans in the construction business take advantage of with regards to hustling the system, of big city municipal construction contracts.

Preface

By adding five to seven workers every week on top of the actual 100 or so crew of workers, providing the city with fabricated names and hours worked to collect payroll for the non-existent employees.

The system was already in place to help in these types of endeavors of check cleaning in New York. This businessman was known as the Jewish Banker, who actually had a bank situated inside his house. After giving The Banker close to $150,000 over several weeks minus the 20% for doing business, I pocketed the rest of the overage as profit.

Understand being a product of my environment I saw this as putting things together in a New York City kind of way and making them work, for me.

In 1995, my cousin took me to the church in Brooklyn with the long lines on Linden Boulevard and there I was introduced to the idea of dealing with the issues of life from God's point of view. It would be three years later in that same church, where I heard a Father's Day message that empowered my Spirit Man to guide my exit from the Big Apple and relocation to Michigan to be a father to my children and continue my personal growth.

Full transparency, The week before I had a visit from Secret Service Men of the US Treasury Dept. because my check game moved up to the big boy oil money and that's all I am going to say about that. There were just too many opportunities for monetary increases, legitimate ones and illegitimate ones. My formative years' neural growth just had a better education of how to process the illegitimate ones faster

And then, I heard the Father's day message, that helped me to mentally conclude that it was time to stop taking a bite out the Big Apple, before the Big Apple took a Big bite out me.

With my focus and desire to be Spirit-led, I returned to school and earned an Automation Engineering Degree. Yes, some would say I took the long road up the rough side of the mountain, but now I pridefully put things together and make designs for positive production.

I became a successful Robotics Engineer, and Prison Ministry Leader with a focus on bringing spiritual consciousness "to those who will" to initiate, the Inner dialogue of the choices we choose to tell ourselves, and the relationship to the consequences that come from the story we just told …ourselves.

I went from "Brooklyn Hustle", to prioritizing money and drug use over family career to being dishonorably discharged from everything, I thought I was going to be.

And this is where the journey begins.

I now have strong bonds with my children and grandchildren, as well as a solid retirement fund that will allow me to live my dreams as an inspiration to those who either want, or don't yet know there is, a better way to live their lives.

Preface

Fig 1

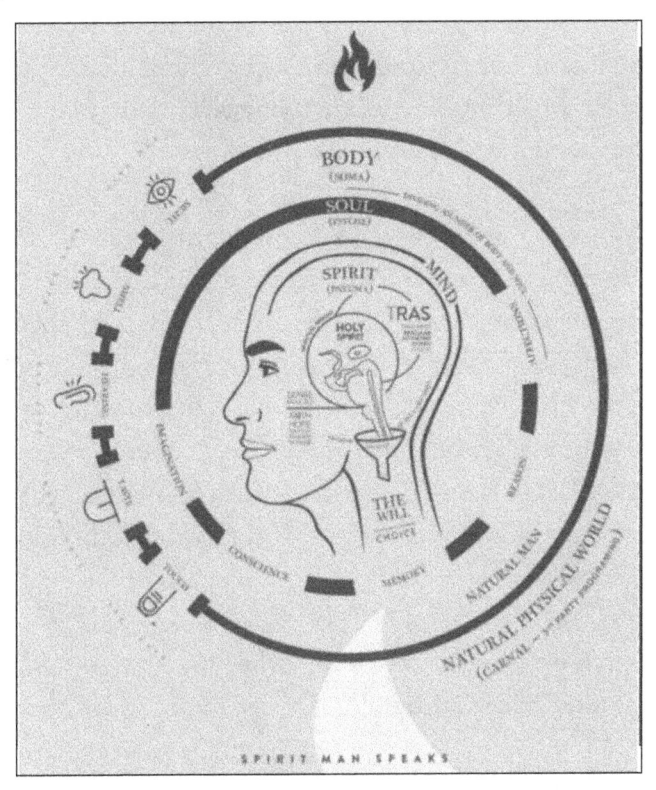

I wrote this book to articulate my thoughts on how to master the human experience. We come into this world with an innate awareness of our spiritual nature. However, throughout our human existence, we are taught and influenced by society's teachings and expectations, which tend to ignore God's intent for the way we live our lives, pushing us further away from our spiritual nature. As a result, our original spiritual nature is not impressed upon our Natural Man, causing us to live a substandard life.

This book is designed to give advice and instruction to help us transition the way we develop our thoughts. It offers guidance to grow from basing our thoughts on the influence of those around us (third party programming) to thoughts based upon a sense of something bigger than ourselves and guided by an unchanging and consistent concept of love.

Throughout this book, we will acknowledge the fight for mental territory that exists between the spiritual self (the "Spirit Man") and the human self (the "Natural Man"). I learned about the battle between the Spirit Man and the Natural Man in "The School of Hard Knocks," not the church. However, when I began attending church later in my life, I came to realize that the Natural Man is miseducated when social influences develop the core curriculum for behavior. This miseducation leads to a "hard knock life."

My church experience caused my belief system to become centered around "God is Love."

To have the most effective human experience, I must allow God, aka Love, to guide all of my thoughts, words, and actions.

If your attitude is, "the way I am living my life and making decisions is not working for me," and there is a deep-down desire for your life to be different and to be more, then this book is for you. While I could be wrong about God being Love, I confidently say to you that if you choose to believe that God is Love and allow love to be your constant guide, your life will radically improve.

— Brother Brooklyn

Table of Contents

Chapter 1 - First Grade 1
Who Are You Going to Listen To?

Chapter 2 - Second Grade 7
Brooklyn Hustle - The Blueprint

Chapter 3 - Third Grade 13
Donkey-Parts on Fire

Chapter 4 - Fourth Grade 19
The Misalignment of Capitalism and
    Human Consumption

Chapter 5 - Fifth Grade 37
The Neural Pathway Shutdown,
    for the "Better Good" Of Santa Claus

Chapter 6 – Sixth Grade 47
TRAS-R (Thalamus Reticular
    Activating System Region)

Chapter 7 - Seventh Grade 97
It's Easier This Way,
-Path of Least Resistance

Chapter 8 - Eighth Grade **105**
Procrastination and Colored Peoples' Time

Chapter 9 - Ninth Grade **141**
Love Sex, and Oneness
What you Did Not Learn in Church

Chapter 10 - Tenth Grade **159**
Zeros and Ones
The Creator made this thing Simple,
    It's Man who complicates It .

Chapter 11 - Eleventh Grade **193**
The Curriculum
Neuroplasticity and The Creator's Design
How to Put off the Old Man and Put on the New Man

Chapter 12 - Twelfth Grade **207**
Stop Tripping in your Flesh and Walk in Your Spirit
How to Get Over on What Society
    Has Shown and Taught You

What makes great literature? A great protagonist and antagonist.

THE PURPOSE OF this book is to stimulate and increase our awareness of God's gift of Free Will, using the measured Power that's in the breath. The Power that made the first human being a living soul. God, the Creator of All Heaven and Earth, has given the only tool necessary to cultivate the neural pathways in the mind that He designed for the body of mankind.

Lessons From the School of Hard Knocks creates the narrative for the Spirit of Man.

This text reveals how the Natural Flesh of Man, or the antagonist, makes it more difficult for the Spirit of Man (protagonist) to reach its goal-- the goal of being their most authentic self, reflecting God's Image and Likeness.

We are born into an antagonistic and hostile environment. A reality shaped by human beings, also born into an antagonistic manner and mentality. This is the reality of the human condition.

Variables from the issues of life are the engaging opportunities for the Natural Flesh of Man. These opportunities stimulate you to acquire more of what's desired. Can the Spirit of Man defeat its adversary by the end of this book, or the book of your life, with the Energy Force we call Love: the book with the Spirit of Man as the protagonist who overcomes the butterfly effects of Adam's antagonistic mental decision?

The Spirit of Man's growth depends on one's ability to identify the neural pathways that are consciously known to be potentially harmful to your future self: the ability to see more clearly and more often where to shut down those mental pathways created by the stories told by the Natural Flesh of Man.

The Journey is about getting knowledge, wisdom and understanding of how to create new neural pathways that are subconsciously known to be good for your future self.

"The Spirit of Man", metaphorically portrayed as the Protagonist of this book. and "The Natural Flesh of Man" metaphorically portrayed as the Antagonist of this book.

The flesh lusts against the spirit and the spirit lusts against the flesh and the two don't like each other.

Pre-Chapter One

As I reflect on being a part of a generation born out of the "cotton-picking South" in the 1960s, many components were significant to my upbringing. My mother instilled in me a Southern respect for our elders and a Northern sense of pride in being free. Mentally speaking, being "North free" was, and still seems to be, a different type of free than being "South Free."

Images and stories of the lynching, killing, raping, indignation, and anger that resulted from the years of being enslaved because of the color of your skin in the United States have not been positive for the mental makeup of the descendants of those enslaved people.

There is only one way to get over the mental PTSD of the horrible effects of that time period in America, and that is through the visualization of our True Image and Likeness.

Then tell yourself the truth: God, the Creator of All Heaven and Earth, has given you a brain that has enough power to know how to never, ever let that type of negative behavior or *"elephant dung"* happen again.

Through the mental subconscious walk of faith, we are able to create new neural pathways that point to the Love of God.

By learning how your brain works, you can learn how to overcome negativity and challenges to achieve more positive outcomes in your life. Your brain can create new neural pathways that lead to love and positive images.

More importantly, we can weaken and shut down old neural pathways that lead to memories and visualizations of disgust and add negative weight to the mental process of manifesting. We should reflect the Love of God in our actions for the world to see.

The journey is about getting knowledge, wisdom, and understanding of how to create new neural pathways that are subconsciously known to be good for your future self

*Lessons From the School of Hard Knock*s is designed to provide the answers and tools you need to overcome a hard-knock life and begin to use Love as the guide and basis for all you do. These ideas will become clearer as you continue on this journey.

Please continue to read. # Knowledge Is Power!

CHAPTER ONE: FIRST GRADE

Who Are You Going to Listen to?

Flatbush Brooklyn circa 1966
PS-269 Elementary School - Kindergarten

REFLECTING BACK ON that first day of kindergarten and the freedom I felt; It was exhilarating. It was one of my first memories of feeling free to do what I wanted. Can you relate to that first feeling of autonomy and freedoms of your parents not being around? In other words, they were not around to tell you what to do.

I recall that first day of school and one of my first memories of disappointment. I recall my subconscious feelings, especially when the teacher was trying to "shut me down" with a kindergarten nap. The storyline I recall amounts to "What the heck lady, who are you to try and tell me what to do?!" Taking a nap was too much like going to bed at night at eight-thirty. It was truly the time of wonderment at that stage in life, and life was just too interesting to let it pass me by due to sleep.

I ended my rationalization process and punctuated it with, "She is not my mom, and I don't have to listen to her." I am sure that increased the weight of self-reasoning concerning what I would do next to at least 51%— that, "I am not taking that kindergarten nap."

My subconscious stimulus for the story, I was telling myself was created from my five years of living on the planet, and limited forethought of outcomes that I had never experienced before. I saw it as the best option to prevent a decrease in something I did not want to lose: to be awake during the daylight hours.

When the rest of my classmates arose from their kindergarten nap, all nice and refreshed, they looked at me and realized I had not been asleep with them. And as I was sitting at my little kindergarten desk defiantly, Mrs. Spellman went to her tall teacher's metal cabinet in the front of the room, and she pulled out a bakery box full of Star-shaped chocolate-covered graham cookies.

After visualizing one of those star-shaped cookies, I remember my mouth watering and imagining how savory it would taste. I remember our wonderful teacher exactly as she began passing them out, as the class slowly returned to their desks after nap time. When she got to my desk, Mrs. Spellman leaned over, whispered in my ear, and said, "Rodney, you are not going to get a cookie because you did not follow the nap time instructions."

If my little mind knew how to curse at the time, I'm sure it would have said, "Well isn't this a b***h America!?" *(RIP Bernie Mac)*

Who Are You Going to Listen to?

As I look back on this very stressful moment in my life, in this public setting, the options raced through my little mind of how not to miss out on these wonderful-looking cookies, cookies that my little young taste buds were crying out for, in the language that little taste buds do at the age of five. Looking at the rest of my classmates enjoying their cookies and milk, I took a mental photograph of the little smiles they all wore on their faces as they munched down on their star-shaped chocolate-covered graham cookies. This was not the outcome I expected.

Reflecting on that day, I now realize that Mrs. Spellman could have embarrassed me a lot worse by telling the entire class exactly why I wasn't getting a cookie. The next day, nap time came again, but my little brain had done some re-figuring the night before. I figured I needed to change how much 'weight' I was putting onto my choices. The option of not obeying my teacher's instructions was lost to the fear that I might miss something. I had a thought for my future self regarding how good that star-shaped chocolate-covered graham cookie would taste when I put it in my mouth. I would follow the instructions from Mrs. Spellman. My little brain thought about this all night long. With my recalculated stance on nap time, I reasoned, then visualized a different outcome and told myself a different story.

This time, when given directions by my teacher, I followed the instructions during naptime, and yes, I got the cookie. My kindergarten teacher used a classical conditioning practice to encourage me to use my free will of choice to change what I thought about nap time; yet nap time, and the cookie, were just vessels for other life lessons. The stimulus for me to take a nap was created by third-party programming (administered in this case by Mrs. Spellman), introducing a different stimulus in my thought process regarding taking a nap.

To The Point

-Word Check: Stimulus
A thing or event that evokes an input into a person's senses and causes a reaction or response. A specific functional reaction in an organ or tissue. (i.e., the sensory receptors found in hearing, touching, and tasting 'the world.')

-Word Check: Receptors
Groups of specialized cells that detect changes in the nervous system's environment (stimuli) and lead to an electrical impulse in response to the stimulus. Sense organs contain groups of receptors that respond to specific stimuli, in which the stimulus becomes a paired response in a subject's mind.

-Fact Check: "Classical conditioning" involves involuntary behavior based on the pairing of stimuli with biologically significant events.

Who Are You Going to Listen To?

The emotional disappointment and the biological fact of not getting to taste a star-shaped chocolate-covered graham cookie created a new neural pathway from my conscious mind to my subconscious mind and rooted a memory. The concept that sometimes you must follow the rules of adults other than your parents was encoded and programmed in my thought process. That day in kindergarten, the taste of the cookie in my mind paired with my decision to listen to the teacher's impending instructions, consequently leading to redirection in my behavior.

In my conscious mind, the stimulus was the cookie. Still, in reality, the subconscious mind did not like the stress of feeling the emotional disappointment and stimulated the brain to develop a plan, not to feel that way. Lesson learned. *Take the nap.* The new equation linked with other neural circuits of thought, rewiring the brain, creating a new boundary of thought, and increasing the reasoning value of doing what I was asked to receive something I wanted.

The reward system was clear. I would often use this equation to motivate and stimulate my thought process that produced the motor functions of what I was going to do and how I would do it. This allows room to create a well-traveled roadway for information to flow freely, back, and forth to the brain, resulting in manifesting into a repeated response that becomes our habits. Societal norms typically tend to place a label on what they see of your 'habits.' It is when we accept those labels, however, that we turn them into our beliefs. In short, once we have the neural network firing the right combinations, we repeat the actions more fluidly. Once we accept those labels, the neural pathway becomes an eight-lane superhighway, with information traveling very quickly. With so many of those repeated actions (now considered our habits), people now identify with and label the behavior as our character. Once you accept the label of that character, we can conclude that your destiny will be what is labeled by the world.

Chapter Two: Second Grade

Brooklyn Hustle – The Blueprint

Tootsie Roll
Cash, Candy, and Compassion

Let's set the memory—passing the store on the corner, every day on the walk home from school. The one that sells the Mike and Ike's, and the Now and Laters. It's back in the day (September 15th, 1967, to be exact) and the jars upfront, of penny candy with Squirrel Nuts, Mary Janes, and Tootsie Rolls, are full and plenty. There were three of us entering the corner store, and we all had money.

In full display, next to the newspapers, there sat the jars full of what had to be over a thousand penny tootsie rolls, triggering my sweet tooth and sending a signal through the already-known neural pathways to my brain. The pathways that stored the memory of how delectable Tootsie Rolls were activated my brain cells to start and figure out how to satisfy the craving by taking a little of this memory and a little of another, or others, and splicing them together into a visualized plan. At that time in my young life, and at a time where everything sweet was not only good, but very good, Tootsie Rolls most certainly fit into that category, where memories were stored in my mind. But, as much as I loved sweets, I had a greater first love for something that was neither candies or treats. That love was for money.

By that time, I had decided in my little seven-year-old head that I liked money, and I liked having money—tucked away in my pocket, nice and warm. Keeping the money was the stimulus for my actions, and from there, the thought calculations began. My brain began communicating via electrical signals, sending and receiving information to each other on how to get over, and keep this money in my pocket.

The store owner wanted my money, and I wanted the goods, but the store owner had the goods, and I had my money. The owner allowed me in his store to get my money, but I wanted his goods *and* my money. I made a plan in my mind of how to make it happen, and the brain quickly identifies exactly where the other parts of the body are. With the hand being quicker than the eye, I paid for a couple, making it easier to 'finger mumble' a few pieces for free. I now know it to be a sign of, and a result of, weakness in man's attention span. I appropriated the candy with the quickness, and the misleading belief and misconception that I was "so fast, the owner didn't see it."

I was consciously doing one thing while subconsciously doing something else. As my conscious mind was leaning up to hand the store owner the money and the candy I wanted to purchase, the subconscious mind coordinated the precise movement necessary for the left arm extremity attached to my five fingers to grab more than a few of those delicious Tootsie Rolls than I had compensated for, and place them in my left front pocket.

The trap had been sprung (where the twisted got twisted), and the antagonistic thought process equated that doing a negative made a positive make sense. This laid the foundation for a mental neural pathway of a thief who steals (the word/label of 'thief' has a well-deserved negatively connotative label, typically based on unfairness to have something without putting forth the equated labor and effort to have it.)

To bring the concept full-circle, it is also one of those ten commandment deals of what you should not do, and for a good reason. Because of the play on God's designed reward neural pathway, a pathway to God will always be the flip for the antagonists of this book. God designed it one way, and man flipped it another. God's way is always simple; however, it is man who complicates it.

It is a lesson of how the antagonist plays the long game in his schemes, with a plethora of them planted in our formative years. The revolution is about developing a curriculum for our youth that teaches the benefits of neuroscience and spirituality by bringing the subconscious authority in the spirit of man to a conscious level. It is the concept of, having done what you 'have always done' to the point of insanity, rewiring the neural pathway to expect change only after a change in behavior occurs.

This can be seen in analyzing the fallen Adam's thought process in the formative years. Trusting above all, the understanding of how to tap into and insert the Love of God at an early age will result in more positivity than ignoring its existence and not doing so. Trusting that the many members, but one body, can rally behind this concept of love and spirituality introduced by the neuroscience, and developed by the studies of how the human brain works with this added irrefutable fact that God's original intent was to manage the body that He designed for His Purpose and Glory.

I was operating under the minimal thought process that I could get a double reward by possessing the candy and not decreasing my money. My God-designed reward neural pathway was now getting a 'double shot,' while I foolishly desired a 'triple shot,' reasoning that "time was money," and I was capitalizing both. The equations of how to accomplish the desire of satisfying my neural reward pathway stemmed from an education in the borough of Brooklyn, NY, and was taken in through my sensory network every day, and all day, from the city that did not sleep. The "Brooklyn Hustle" is about going forth and multiplying those memories made of what I saw, heard, and felt, coupled with the feelings that I experienced from having less than, and what I could do to change it. On one hand, I could follow the script that society had written for me to follow, and at that age, it was hard not to fall into the trap of flipping God's original design for the reward pathway. Yet on the other hand, I was not aware of one's ability to change your own neural pathways.

What was I letting grow in my subconscious mind? Negative concepts of what your five senses take in response to the cycle of negative, produced by the brain, with the same theory applied to positive thoughts. The negative that I was inflicting on the store owner never even crossed my mind, and I wanted to keep those seven cents in my pocket.

There was something to be imprinted in my neural pathways on passing Godly Love to and from my neighbor or community. Simply put, the blueprint is stimulated by opportunities to increase or stimulated by opportunities to prevent decrease. By correctly dividing the words, labels, and phrases that enable you to envision the end from the beginning, you will possess the ability to acknowledge 'where the twisted gets twisted at' (or the point of confusion), based on these two serious, and most important inputs.

To the Point

-**Fact Check:** When a thought is produced, various neurons passing an electrical signal between one another are responsible for that thought. Any time two cells communicate, the brain creates a connection for any future communication.

-**Fact Check:** The brain continually rewires itself by adding and removing connections to the Neural Pathway. This is how we learn new information. Positive means good, or the opposite of negative, and negative means bad, or the opposite of positive.

Chapter Three: Grade Three

"Donkey-Parts" on Fire

Where Does the Pain Go, When You Learn.

"You can be whatever you want to be, because you are a child of God."

These were the most influential words my mother programmed in me. The second-best thing my mother programmed in me was discipline, which created a path in my thought process.

I now wonder, did my mom know she was creating a neural pathway in my brain? One of the most integral parts of this pathway were the lessons stored in my 'memory box' labeled, "I don't want my donkey-parts to hurt anymore!" This is the box from which I could recall and retrieve memories--- memories of what that thick leather belt that hung on the back of my mother's closet door felt like colliding against my thin skin (and that's where this memory begins).

At the initial onslaught of nociperception (the sensory nervous system's response to harmful stimuli), I experienced a severe case of neuropathic pain, and the damage born from the belt's rhythmic song being played on my donkey-parts got the attention of the sensory neurons, quick. *Whip, whip, whip!* The sound released impulses along the axons involved in the pain-signaling pathways, sending the neurons traveling up the spinal cord in the nervous system. Nociperception fulfilled its designed function of detecting a change in an organism's physical or chemical structure (in this case, my donkey-parts) while emitting white lightning through billions of neurons, which make up my nervous system. I think most can testify that the sensitivity of the tissue in the immediate area of the backside is at a higher level than other parts of the body.

This five-alarm fire signal traveled fast along the tracts of a specifically created neural pathway, flowing information from one neuron to another at a junction called a *synapse*. As the neurotransmission continued between presynaptic and postsynaptic endings in my backside, it was transmitted loud and clear that my donkey-parts were under duress. This fired a sequenced impulse so precise that it connected only to a specific neuron, and that neuron, in turn, the same to another. These neurons were now on fire and communicating with each other via electrical events called 'action potentials,' making their way on their own highway to the brain.

The message was the same as it was from the initial point of pain to my thought processing, dissecting what to do about it. The ability of an organ to respond to external stimuli is called sensitivity. It is when those signals reach the brain. My brain immediately went to work on how to stop the pain. One of my brain's first thoughts was to try communication by sending signals to the neurons in the vocal cords to articulate the following words as fast and as many times as humanly possible: "Momma, Momma, please, I'm sorry Momma!" "I won't do it again, Momma, I promise. I promise, Momma, I won't do it again!"

Unfortunately, this bit of attempted communication is not being received on the receiving end, because uncontrollable screaming can limit how much air we can take into our lungs, which then makes the syllables very short on pronunciation, diminishing your very sincere apology to an unintelligible noise to your mother's ears.

-**Core Consciousness:** This emergent process occurs when an organism becomes consciously aware of feelings associated with changes occurring to its internal bodily state; it is able to recognize that its thoughts are it's own and that they are formulated in its own perspective in the present time.

Understand, the brain does not like to be under stress, so mine was dealing with this problem with an 'all hands-on deck' mentality and sending signals to different neurons and other components of the body to assist in stopping the pain. The brain will engage the motor-sensory, transmitting signals to your one free hand, warning it to block the incoming blows to your donkey-parts like an offensive lineman in the NFL. Through all of this, your brain continues to process over 10 million bits of information per second.

Imagine the mental processing required to activate the body and imagine the amount of electrical energy required by the brain to fire the correct sequence of neurons. Then do the calculations on how to stop Momma in full "whoop donkey-parts" mode!

First, the nerve endings in my backside sent electrical impulses along axons to my spinal cord (magenta pathway). Those incoming axons form a synapse, or connection, to join neurons that build pathways up to the brain. The neurons that travel up the spinal cord then form synapses with neurons in the thalamus, which is a part of the midbrain, or magenta circle. The thalamus reticular activating system region manages this incoming information, with a series of changes in our mental activity until we reach a final conclusion. That conclusion becomes the action that we choose to execute. The thalamus sends this agreement of signals back down the spinal cord, which directs the motor neurons to the portion of our bodies that need to take action.

In this instance, the brain fired the correct sequence of stimuli that produced neuronal transfer through synaptic gaps. Subsequently, the brain then rationalizes, "I could just get up and run, but where will I run to if I get up and run?" "Where would my food come from, and what shelter would I have over my head?" Ultimately, the brain reasoned with, "I guess this is just one whooping that my donkey-parts are going to have to take because the alternative will be a bigger negative than the one that I am facing right now."

I was taught similar mental lessons by my grandfather during my summer vacations on his sharecropping farm in North Carolina.

We often sat at the feet of our elder relatives, and they shaped my formative years with lessons that still ground me today. As

"Donkey-Parts" on Fire

a child, I was oblivious to the hardship of that time. It completely flew over my head. I believe the largest explanation for that would be the elders, the family, the cousins; everyone was so connected in love— the love of God, the land, and the family. Summers away on sharecropped farmland can shape a man in his youth and redirect the conditioning inside.

I think back on my childhood and summers spent down South and we were told to go get a switch off a tree to cause bodily harm to your self.

Can you image how many corrections and the number of times the young thought process had to correct the bodily movements to follow instruction to the question, so that it made sense to do it.

Why am I going to get something to beat my own "donkey part with?

Those lessons are valuable today because I now understand the type of neural pathways that early childhood experiences can create. Sometimes the message was for my ears, reminding me to pay attention. Countless other times, the messages were for my eyes, serving as a warning to look both ways. Then, there were other times when the message was for my nose, serving as a reminder to keep it out of grown folks' business.

This is more than sufficient evidence to demonstrate how much power your brain has over your body. The brain could have sent a signal that told my feet to get up and run, but through all the pain and anticipation of pain to come, the brain calculates the best algorithm and plans to return things back to normal, in a safe way for the whole body.

This could probably best explain the notion of staying still and not running when we were told to "Pull your britches down!" along with, "You best not run!" when in trouble back in the day.

Looking back, I don't believe our young minds ever got to the point of even fathoming what would happen if we did not follow the dreaded instructions.

That said, peace be still/feet be still, and not run.

It's very important to bring this subconscious fact into conscious thought. The brain that God made has the power to control what the body does under the greatest duress and will always do what it thinks is best-- to survive and get back to running the other 100 billion neurons/nerve cells in the body.

"The body without the spirit is a pile of dirt." - SpiritManSpeaks.com

Chapter Four: Grade Four

The Misalignment of Capitalism and Human Consumption

The Peculiar Institution, Who You Mad At Bro!

"The institution of slavery was, for a quarter of a millennium, the conversion of human beings into currency, into machines who existed solely for the profit of their owners, to be worked as long as the owners desired, who had no rights over their bodies or loved ones."
-Isabel Wilkerson "Caste: The Origins of Our Discontents"

The nurturing, development, and cultivation components of my mental thought process were handed down to me through third-party programming, aka the lessons my mother taught me. My mother's mind had closed down a number of neural pathways from her programmed youth, as she programmed her new thought process into me. This part of the third-party programming that I received from my mother was in full force.

Each experience with her, stored in my mind, would assist in shaping my reality. My mother's helping hand guided, protected, and pressured me in the direction of her definition of what I needed to be. Her programming from childhood closed down some integral neural pathways defining my current state. As I encoded these lessons and memories, they were to become a part of my foundational truth of beliefs in creating the neurology of my master program, The thoughts of myself, based on every memory that I have ever had, and creating the neural circuitry that makes me, me. We are who we are, and we are who we have been. It is the same neural circuitry that makes you, you. We are who we will always be, an energy force of Love.

Capitalism and Human Consumption

"Neither slavery nor involuntary servitude, except as a punishment for crime whereof the party shall have been duly convicted, shall exist within the United States, or any place subject to their jurisdiction."
-United States Constitution, 13th Amendment

Circa 1550, slaves were depicted as objects of Conspicuous Consumption (to be used), with slavery itself becoming inseparable from consumerism.

Was the Curse of Ham twisted to keep slaves in their place? Why was Canaan dropped from the story, with Ham and his descendants made to be African? Research the location of present-day Canaan today. What does the geographical location suggest? If God cursed the Canaanites, how common is the belief that one is a descendant from a neighboring land, like Syria, Lebanon, or Jordan? Did Eve originate in Africa? Which region?

Knowing something in part and not in full truth is nearly the same as believing a lie. What if this lie has shaped your complete belief system since childhood? It does not change what the truth is. You will react, respond, and surround yourself with what you believe. Every day we experience the consequences (good or bad) from those responses and their subsequent actions.

Records show that between 1865 and 1920, after the Civil War, hundreds of thousands of African-Americans were enslaved again in an illegal and abusive holding of freed negroes, specifically in the South, called "peonage." In mass numbers in the deep South, African-American. men and women are "leased" to coal and iron mines, plantations, brick factories, and other harsh work environments and conditions. There is a loss of empathy for understanding the thought process with these memories from our history.

I was not taught these truths in school, and I doubt that many other people were taught the reality of American history either. Without mindfulness of the Black man's history in America, a disconnect is likely to exist. Functioning in a society where laws and policies exist showing bias will eventually shape the conditioning of a person. Knowing this history is empowering, yet simultaneously tragic. I bring these words of things in the past up because of their effect on things in the present. Words possess tremendous power!

It was the threat of knowing that if I did not do what the third-party programmer required, it would be necessary to do physical harm to the body. The brain is designed to take the path of least resistance, and it was not trying to hear all that. It functions as it does to manifest the appropriate motor-sensory skills of the body to keep it safe, and as safe as deemed necessary for further living.

The Rules of Engagement for the Uppity Black Man

Can you imagine the mental stress of having to answer one of the worst questions that could be asked in the middle of the night in 1958, America?

"What are you, some uppity negro?"

That question holds an implied threat, that if you are this type of African-American., "one who lacks obedience," then you are in danger. This social control was written into law and exercised by officers of the law, and the government, delegated to administer the rules.

The very institution that honored our freedom, sought to bind our citizenship to a degraded 'free' existence that is always under the thumb of the once- chattel owner. The theory that one must now treat African-Americans differently with a separate set of rules to steer them back in the place of mental submission was exercised by direct violence.

Many political, social, and judicial advancements for African-Americans were halted due to direct persecution. The perceived

The Misalignment of Capitalism and Human Consumption

threat caused generations of damage in the recollections and stories of memories handed down to the descendants of the "peculiar institution."

By having to recall from your memories the handed-down stories of Black men hanging from trees as a result of their "uppity-ness," we ignite a fear response. The Black man's brain says to his body language, "Neck, hang my head down." So, I don't stress out from the thought of making another human feel intimidated or that their superiority is being challenged. It is mind-blowing how these shared experiences connect us and give us the ability to empower and support each other. This is more evidence that we are 99.9% the same.

Owning another human body was an inalienable right in America if you had the money and wealth, which means by society's standards and norms, you had the power as well. The everyday business owner, politicians, and even clergy, all benefitted from chattel slavery in America in some form. Take God's Sovereign Power over His creation, trying to be duplicated by crazy-a** Adam's "selfish" thought process.

This led to chattel slavery, more self-glorification, and ideas of racial superiority. Is the love of power truly the root of all evil? Or is the love of money the root of all evil?

Although slavery produced a large amount of money and was a major part of the economy, its popularity truly stemmed from the desire to have power. The thought process that compelled Adam to make that selfish choice is a part of all humanity today.

The fallen mind, at its worst, will use power and money to take shortcuts in his desire, helping him to get rewards.

To put it plainly, you do always have a choice; you have a choice to live or a choice to die because another human being wanted to have sovereign authority over you so bad that he would kill.

This authority went to great lengths to constantly display power over the Black body through cruel and unusual punishments. The practices were meant to serve as an example, of mental manipulation to control the population mentally. Bodies were maimed, and men were tortured and killed in a multitude of vicious and hateful ways at the hands of this "sovereign authority"

Not all, but a lot of white Americans acknowledge and show the empathy our nation needs to heal.

Now I understand why, when talking to the descendants of those enslavers, they are quick to distance themselves from that time period; because of the realization of the visualized, images depicted of a human body going through what the black human body had to go through is definitely is not a situation I would want my own human body to have to go through.

However, this social system of slavery has generational implications we must acknowledge. On the other side of that, it's the reason for racist words that exist today and the labels created by societal norms that are to be used when you feel you don't have sovereign authority over a black person's life. These slurs and ideals still exist. People are called the 'N-word' for two reasons: the label that societal norms assign to it, and how it aligns with your belief system. When we look at the words used during slavery, after slavery, and even today, we see that they are all those types of phrases that elicit a response. When these words go into our ears, and if a neural pathway exists in the mind that says we are no good, or the belief that we are something less than, it is very possible that when those words connect with that neural pathway, it will add some weight to the decision-making process concerning what you are getting ready to do, next. Because there is always a what's next.

That same human brain that had to deal with the aftermath of the peculiar institution is a descendant of the same sequence of neurons through sensory networks that made the slave brain bow down for survival. The power of the brain to control the body is real: that same brain that told your body parts not to run. The brain was designed to take the path of least resistance back to normal and safe, so it took those licks to get back to safety. That same brain of the slave that told his body parts not to run, recalling memories from the peculiar institution.

The memories of the tarring and feathering, the memories of the horses pulling a human apart. The memories of families being split apart, designed to break the unity, hearts, and spirits of the slaves and their thought processes. The enslaved person's thought process was only to have limited mental activity, similar to what we would assume would happen to a mule with blinders on. Our feelings, motives, and decisions are powerfully influenced by our past experiences and stored in the subconscious mind, including our wishes and desires, traumatic memories, and painful emotions that have been repressed. So, let us un-repress them now for once and for all.

Who Are You Mad At, Bro?

"I don't want to get whipped anymore, so I'm going to pick this cotton, and thank you for the room and board while picking this cotton... Thank you for the lesson in capitalism and man's mental ability to manipulate other human beings for capital gain."

As the Spirit of Man was in our forefathers, this period required us to rebuild a new neural pathway to God. People were not allowed to read and faced punishment if they were caught learning how. Think about what you are capable of now that you are allowed to read.

The brain likes the path of least resistance. This is one of those little-known facts that is a big deal in regard to the truth. A slave-whipping and the mental effects that experience would undoubtedly have on a human must have been immense, along with its impact on those witnessing. Constantly being conditioned to accept the fact that if I don't change my behavior, death may be the only outcome right now. The internal crossroad of where you will have to choose between two inputs of thoughts; one that wants to give way to the path of least resistance, and the other desiring to create a different pathway, which will alter the current path of least resistance.

It subconsciously redirects and focuses on preventing a decrease and never getting caught out like that again, vulnerable to letting a negative situation transpire, where your survival is based on the next choice you make. It's logical to see how the "master chattel owner" programmed a belief system to condition our forefathers. This conditioning could be handed down generationally and was constructed to impact our mental health and memories so that we would never forget the generational stories that are told recalling American slavery. The peculiar institution is embedded in the subconscious of our youth today.

At times, the limiting mindset from that peculiar institution gets handed down in those generational stories with words that force you to visualize and create trauma in the human minds of most of the descendants of the system. The rEVOLution is about bringing understanding to conscious thought. The whole deal is to expose where the twisted gets twisted so that our youth only have to hit their heads once or twice, learning the lessons. Then, they will possess the ability to recognize the power of words and their effects.

Less than 70 years ago, my colored donkey-parts could have been hanging from a tree--- dangling lifelessly and swinging back and forth in the slight force of mother nature's wind, the weight of the body, succumbing to the law of gravity. Bound by the rope around my neck, with the other end tied to the law of stability, a big tree branch, overwhelmingly taller than me. The only thought left in the brain is the notion of saving its body and removing the rope from around its neck. I would be sending motor-sensory signals to my arms and hands to free my body from the danger, yet I'd get a return message declaring that's not going to happen. Contrarily, these other humans knew that, and they tied your arms behind your back. Therefore, the brain now thinks and believes it has only one trick left to escape, escape to air. Bypassing conscious understanding again, that the law of gravity continued to do what laws do, my windpipe collapses …I can't breathe…fade to Black.

The power of the mind allows it to recall clearly, as if these events just occurred yesterday, because the brain has attached the emotion of the event to its visual image. They are now encoded together. This type of subconscious behavior needs to be brought to the conscious conversation.

Subliminal Negative Consequence of the Peculiar Institution
"Where the Twisted Gets Twisted"

To the Point

Case Study #1
When a Black person says to himself, "It is, what it is," vs. when a White person says the same, what are the thoughts of the consequence within this conclusion? What is visualized by each brain when we say these taught words and phrases to ourselves? The phrase "It is, what it is," came on the scene in 1949 from the thought process of a man dealing with the repercussions of the harshness of Nebraska's lands.

"It's cold out here, the land is hard, but I've got no other choice but to deal with these harsh conditions and survive and keep it moving towards my goal. It is what it is, without apology."

While we've introduced that phrase as a societal norm, it is also used as a motivational tool. Corporate says it is what it is by leading with the attitude of "let's focus on how to get around that, with the resources that we have," or, "we have no other choice." The military says the same, in the same regard, as well as the government.

The Misalignment of Capitalism and Human Consumption

What must be accepted? What must be accepted is what the General tells you to do next: possibly demanding you to go and sacrifice your life. What must be accepted is what the corporate boss will direct you to do next: which could consist of laying off half of the employees. What must be accepted is what choice the government will tell you to make next: pay your taxes.

It is what it is.

What will a man do, or put in a box labeled with that phrase, and have no morality or acceptable excuse for the manifestation of his action to survive? When he feels he has no other choice but to use the resources available to him and listen to his own understanding of the current moment to survive, essentially, he will enter "MacGyver Mode," with all things around him being useful, if the mind can think of a way to use them. Go for yours, and what you desire for your life, the best way you know how.

"It is, what it is" was intended to be visualized as an acceptance of whatever the issues of life throw at you, and instructions on how to deal with the circumstance the best way you know how, ceasing it from stopping your forward-moving and progress. It is an expression of words used to characterize a frustrating or challenging situation or issue of life, that the human mind believes cannot be changed and must be accepted. Subconscious behavior being brought to a conscious conversation, actualized, is the realization of taught formulas and equations of words, labels, and phrases from the left-field mind of Adam. It would be practically impossible to possess a consciousness that your 'path of least resistance' is corrupt, and have a different outcome, conclusion, or result.

The taught response of giving up creativity and accepting the consequential conditions in a moment or situation--- the neural networks of a African American person versus the neural networks of the path of least resistance of a European American person. How many Black human minds versus White human minds visualize this "taught phrase"? I want to continue to subdue all the creatures of the Earth.

The final conclusion of acceptance taught to our subconscious behavior, which will be manifested in the members of our bodies, subconsciously, with full authority to bypass conscious thought, given certain protocols of influence from the other human mind.

"...Fear in the brain will produce a path of least resistance."

The human mind will figure out what to say, with the best articulation for visual interpretation and clarification. The "hear, speak, see no evil" theory is present because the mind is more interested in preventing a decrease for its future self. The human mind is always interested in how to control increase and decrease.

Can you imagine the conversation between two masters, with one explaining to the other how to prevent a decrease with their slaves? Two human beings, sharing thoughts of how to multiply, increase, replenish and subdue all the creatures of the Earth. Both enslavers understanding the mental effects of memory, and each understanding the purpose of branding an image of death in the slaves' minds that will also be taught to their offspring.

Visualize a baby elephant, unable to pull a stake out of the ground since infancy, with no concept of what would happen if he was to pull that same stake out as an adult elephant. As a baby elephant, and because of his baby elephant mentality, he was not yet fed any new directions on purpose. American slavery was, in many ways, a study on the motivation of money, resulting from an innate desire to subdue, have subjection, and sovereign authority sovereign authority over another human mind.

Who you mad at Bro?
Woodrow Wilson. When he became the President of the United States, one of the things that happened under his watch and leadership was the rebirth of the KKK. In fact, Woodrow Wilson showed the very first ever film in the White House, and his choice was The Birth of a Nation: the recruiting film for the KKK.

Who you mad at Bro?
Woodrow Wilson, whose picture should be in Webster's dictionary next to the words "Karen's and Ken's Great great great Grandfather". Also from a subconscious behavior point of view, this is what happens when you are taught something that is not true, and you believe it. Woodrow Wilson was a leading professor of United States History, at Princeton University. When he wrote his five-volume work, The Chronological History of America from the beginning to the then present, it was believed to be the most scholarly work ever done in a history book series, and became the United States History textbook for colleges across the United States.

Who you mad at Bro?
Because Woodrow Wilson's "history" removed all the "African" American heroes, most Americans have never heard of the black guy who crossed the Delaware River on the boat with George

Washington also known as Prince Whipple; or the black guy from the Revolutionary war # James Armistead, the black spy who helped Washington capture Cornwallis, sealing the American victory in the Revolutionary War using the valuable information given to Washington about Cornwallis' location at Yorktown.

Who you mad at Bro?
Woodrow Wilson was academically and scientifically accepted, as presenting the truth.

Who you mad at Bro?
What about Charles Darwin? In his early book, the Origin of the Species, with its blatant, yet seldom-cited subtitle: The Preservation of Favored Races, Darwin institutionalized the words, labels, and phrases that continue to help visualize a lie: the lighter your skin the more evolved you are, and the darker your skin, the less evolved, or the more Neanderthal you are.

Who you mad at, Bro?
Men like these, whose popularized misinformation became the only information passed on to the larger population, or the people who were raised, educated, and socialized being told that this misinformation was actually the true information, and they were offered no sources of alternative information? #Alternative Facts

After reading this chapter, one must ask some comprehensive questions: Does it make sense to consider some things differently? What are the lasting effects of words, labels, and phrases that we say to ourselves or allow to be said to us?

In the African-American mind, the descendant of the peculiar institution, you will see how it is at best. Creating new neural pathways in your mind would consist of using the principles of everything you have ever heard about visualization and visualizing what it would look like to make the right choice at the right time. This would also involve visualizing the outcome that produces love, not hate, and an image of yourself that is most authentic to who you claim yourself to be (man of God, woman of God, Kings, and Queens of the Most High).

By visualizing what it would look like to LOVE.
By visualizing what it would look like not to hate.
By visualizing what it would look like to make up, build up, and not tear down with words, labels, and phrases of low vibration.
By visualizing what it would look like to have a likeness to the Creator of all Heaven and Earth.
By visualizing what it would look like to portray the image of the Creator of all Heaven and Earth and manifest it through your living body for the world to see.

**This is how you get over slavery in America.,
This how you overcome all of man's inhumanity to man**

Mentality. This is not to cause confusion, but to expose where the confusion lies by understanding what is happening in your brain on a subconscious level. It is to shine a light on the areas where the author of confusion does its thing, with words, labels, and phrases. The question becomes, should you continue to take the path of least resistance?

The path that produces the mentality of limitations and reduces you to what societal norms expect: limitation.

As a man thinketh, so is he. The human mind has to have consciousness of "a thing" before producing "a thing."

The foundation for any equation possible to the human mind will always be based on positive and negative numerical slides.

For instance, most people don't buy stock considering the potential decrease. It is considered so heavily and weighted that it prevents them from buying the stock for the potential increase.

Most human beings don't conform or buy into a new belief system unless the potential increase is better than the potential decrease of their current belief system.

When a spiritually conscious person is about to get into an accident, and subconsciously taught behavior kicks in and bypasses the slower thinking conscious thought process, a more religious or spiritually conscious person would not respond by screaming incoherently or wielding profanity and obscenities.

The subconscious action that has replaced what was taught by societal norms will come from the renewed and rewired mind of the spiritually conscious person, with full authority to bypass conscious thought, given certain protocols of circumstance to cry out to a higher power, God, for help. I'd rather call out "Jesus," and receive the expected actions of that belief system, than call out "Oh Sh*t!" and receive the expected actions from that belief system. This is humorous when we look at how simple God has created things, and then we realize how complicated man has made it for his own selfish greed.

Many people are not aware of the fact that many have tried to forgive but can't seem to. Basically, you will not be able to forgive until you separate the emotion from the event. This is one of the places where the twisted gets twisted, in our efforts to love and forgive.

As a spiritual being mastering the human emotions it is my conclusion that the biggest variable between what we ought to do, and what we end up doing (right vs wrong) is rooted in emotional response.

Chapter Five: Fifth Grade

Shutting Down Neural Pathways For the "Better Good" of Santa Claus

THE SUBCONSCIOUS MIND is nothing more than the "neural pathways" that have been established in your brain as a result of your past beliefs and conditioning. During your unconscious existence, when you weren't aware enough, you wound up imbibing a host of limiting beliefs, negative conditioning, and misguided perceptions about life, which were taught by the adults we relied on to tell us the truth, and we believed them, until we didn't. A parent will "fib" to their children, if they believe that it will make their child happy.

As I understand, my mother was doing the things that her parents taught her, (the funny, yet not-so-funny thing is, they were given presents of walnuts, pecans, candy canes, and single dollar bills) including the tradition of Christmas. At that time, the Christ in Christmas was far out in front of Santa Claus.

Parents from the 1950's era, born to sharecropping Negroes of the time, were not making any allusions to Santa Claus, and the giving of Christmas was from love, which is what was handed down.

When my mom moved to New York, and made some New York big city money, she also made some new neural pathways evoked by the new environment, one of them being the giving of gifts to me, from Santa, to express her love to me. This was then paired with the emotion of love for me, and the giving that God gave in His only begotten Son—Amen!

Love really is for free, and you can't get around that; and this was an energy transference of love exchanged through the gift of giving.

Parents often use a number of tools to shape children into wholesome beings, while simultaneously benefiting their parental reward system.

She knew that these "rewards" would make me happy, and that they satisfied my reward-driven *neural* pathway. At the same time, the love she was giving away satisfied her dopamine receptors.

As you could probably imagine, that poses a problem to consumerism and how to capitalize on it So, in the name of capitalism, the grand scheme was to flip the love thing by mentally manipulating it, and attaching it to the instant gratification that comes from giving, subconsciously adding credit cards to fund the scheme, and do it again eleven months later.

Shutting Down Neural Pathways For the "Better Good" of Santa Claus

The Santa ploy also gives leverage to parents—a fact that I figured out early on.

My mother used every bit of leverage that came, from my believing, He knows when I'm sleeping? *What?!* He knows when I'm awake? (What*!* x2) ...He knows if I have been bad... or good!

So, I would be good, and it was in my heart to be good, for the whole year.

Why? Primarily because I did not want to be put on his naughty list.

As I look back the "small white lie" #fib and story of Santa and the reindeer with the red nose, was a good trick, by way of mental manipulation, as it was beneficial in building my behavior, and it provided an incentive for me to control my behavior', in time, building foresight increasing my control over my behavior

This seemed like very heavy stuff because I realized that a lot of the things, I wanted to do were on this Santa Claus dude's naughty list, but I did not, because of the "reward system" conditioning. was created by the concept of *"the naughty list."* the believed threat of *"you better watch out, you better not cry... I'm telling you why... Santa Claus is coming to town!"*, and he's checking his list, *to see who has been naughty or nice.*

It was programmed into me that, *"He knows if I have been good or not, so I better be good for goodness's sake."* I love my Momma, and she came from the *"love isn't love until you give it away"* school, and I am so sure she never gave a thought to the subconscious behaviors that are influenced by false variables of information.

{The neural pathway to the "Reward" had a fake variable" that I believed that I had to use in the equations, the grammatical equations in the story I would tell myself.

The fake variable of information giving, with which I could create a new neural pathway which would produce variable body movements that were fake, that eventually would produce fake productions.

Twenty years of robotic movements, I know what I am talking about, and I say this in the most positive way that fits this next statement.

Older people are sort of given the privilege to say what they want, within reason.

This is within reason, when I make the statement, "I know what I am talking about" and is a subconscious sore point, to the many times it was pre-perceived that I did not know what I was talking about to the same degree that others in the room. At first the "subtle snubs" just made me work twice as hard to prove greater was in me than the negativity of the world,

But it was in my belief system that wisdom was the principle thing, therefore I seek to get wisdom but with all I was getting I got an understanding, of what was going on with regards to subconscious behaviors of the many places I would be the lone

Shutting Down Neural Pathways For the "Better Good" of Santa Claus

African American Engineer. I had to prove, that I knew what I was talking about, because of the newness to the neural pathway of my European American Engineers belief system, that they themselves did not have a neural pathway created to reference from with regards to an African American Automation Engineer, in the Aerospace Industry. In my understanding, they knew no better, and I asked my Father to forgive them, for some things that could be considered unfair, for in my heart I truly believed they knew not, what they are doing; because of understanding of concepts of this book, knowing about the neural subconscious pathways that are connected to confirmation bias that works in every human brain on the planet.

The Antagonist in the story of my life at that time thought he was keeping something from me, based on his belief system, (messed up as it was), but in reality, the Real Reality, they were pushing me towards something … the Reality of Gods Plan.

They really knew no better, so I forgave them, until they could do better, and they did, above and beyond anything that I could have imagined. They gave me the keys to a car and keys to a apartment, and keys to hotels in the United States and keys to Overseas hotels and keys to a "Corporate Credit Card"

Just in case you are keeping count Christ went down below confronted the Antagonist himself and said "Keys Please" came back up with All the Power. (spiritmanspeaks.com)

Growing up in the inner city, with no other neural pathway to reference from except what was given to create the new neural pathways from, you are like my European American Engineer associates, with pre-conceived doubt, in your belief system, with regards to what you have never done before.

The saying, now that you know better, you can do better, because for real "We Gotta Do Better" (RHJ). Now we have a better way to do it. #LFTSOHKS

The Better Good

At first, I thought my mother was upset with my older cousin for spoiling the so-called magic of Santa, but I know now that there was much more behind her emotion of being upset. I knew some information, that she might not have wanted introduced to my neural network yet,

We were stimulated by an opportunity for an increase by the character Santa Claus because we were taught to be stimulated by him with regards to the reward system.

Eventually, we believed the other words, labels and phrases that came from our older cousin, that said Santa Claus was not real.

The unique electrical pulse that had been fired the same way that used to bring excitement, was fired the way it had always been when we heard words, or visualized what Santa Claus meant to us.

Now, because of this added information from the older cousin about Santa Claus, and our belief in what he said, it changed how we added things up to get a visual image of Santa Claus, and the stimulation for an increase.

Shutting Down Neural Pathways For the "Better Good" of Santa Claus

Long story short, Santa Claus got kicked out of our belief systems, and we eliminated an unused neural pathway and replaced it with something else.

We allowed some information in, and kept some information out,

I now possessed a new neural pathway that says it is alright to lie if your thought process equates you will be doing a better good.

It was the subconscious behavior that produced conscious actions of trying to make money, by any means necessary, and not telling my mother the whole truth of where I was getting my money from.

If you are still believing in the things of your youth, like Santa Claus, or that you are going to illegally sell enough drugs to buy you and your mother a house, let me be the older cousin to tell you.

Stop believing the hype that was taught to you about how you are going to make a lot of money, the way your surroundings may have shown you. Understand that it is designed to show you this way on purpose, and I conclude that, the design is broken now, and America and the world are *woke* to the "Santa Claus" method of mental manipulation.

The fake fibs of societal norms are the variables that we add in to our grammatical equations of the words labels and phrases we choose to tell ourselves in, the story we tell ourselves. (Read twice for repetition for creating new neural pathways for the truth to be inserted.

True success in life is not coming from a man, made up to mentally manipulate my mind. It is coming from my mind, made up to overcome man's manipulation.

Let me also be the older cousin to share his knowledge, wisdom, and understanding of how to get over on what was taught to us in our formative years, which doesn't add up to a hill of beans, or really anything at all.

The unique electrical pulse that was fired when we thought about what Santa meant to us and the excitement we felt was now encoded with information about the true identity of Santa.

As a result, a new pathway was formed that bypassed the pleasure receptors formerly attached to the label Santa Claus.

Without a PhD in Neuroscience, we can understand the simplicity of this subconscious behavior, because it is how we learn every day.

In chapter one of this book, I shared the story of how the pathway that was created because of my fear of missing out, was replaced with the star-shaped chocolate-covered graham cookie, which in itself became the new, and more heightened fear of missing out.

In relation to Santa, in my six-year-old mind, this newly adjusted neural pathway carried a new equation to my subconscious that concluded that Santa Claus was not real, but the naughty list, was.

Shutting Down Neural Pathways For the "Better Good" of Santa Claus

The concept of the naughty list didn't change, just my beliefs around the concept. My mom was Santa, and everyone within snitching distance was an elf used to contribute to the list.

-Word Check: Foresight

The ability to see what will or might happen in the future.

The ability to visualize the end from the beginning. The ability to predict what will or might happen with regards to what is best for your future self.

Neural pathways connected to the subconscious brain analyze data, and figure most of the positive and negative scenarios, to come up with the most logical equation to determine what's next, and the action you will take in response to that expectation.

The foresight of understanding consequences of the future is a good neural pathway to develop early. There were times when I really wanted to reach over to the desk next to mine and snatch little Bobby's star-shaped chocolate covered graham cookie. But I didn't because I knew it would put me on the naughty list. More importantly, I didn't want another episode of "donkey-parts on fire

Master the Human Experience

If the antagonist of this book cannot make you mad enough to sin with whatever words, labels, and phrases that come into your external senses, and are processed through your neural network, you may be able to disconnect the visual image from the past emotions that it was attached to.

By simply acknowledging the operating procedures of the brain and what it is capable of, you are creating a neural pathway, by repetition and replication, of the things you just read

If you repeat it enough to yourself, you will assist in strengthening the synapses between these new neural pathways.

-Word Check: Synapses

Synapses are the communication links between neurons. They build the neural pathways to be used for all of our thoughts, behaviors, actions and feelings.

In sum, it is the electrical impulses being sent back and forth that use force and energy to create a *'deeper rut into the path,'* every time. The electrical impulses jump from synapses to synapses, firing the same exact way, every time. The exact same frequency every time. Increasing the speed of information that ultimately becomes part of your reaction time into another human being's thoughts, behaviors, actions, and feelings.

The function of your subconscious mind is to *store and retrieve data*. Its job is to ensure that you respond exactly the way you are programmed. Your subconscious mind makes everything you say and do fit a pattern consistent with your self-concept, and your "master program," or who you believe yourself to be.

Now my friend, the plot thickens...

Chapter Six: Sixth Grade

TRAS-R Thalamus Reticular Activating System Region

The Space Between Electrical Thoughts, Mechanical Energy and The Members of our Bodies

Any word, label, or phrase that we can use verbally will always be a word, label, or phrase that another human being has visualized that word is or should be.

God, the Creator of All Heaven and Earth, is my visualized word/label, worthy of starting a conversation.

As a free-willed human being, the spiritual element of me could not answer the question as to why I should think that my free will to choose and think a certain way should be another person's way to use their free will to choose and think.

The spiritual element of my intellect cannot deny the natural fact that if I were born in another part of the world, my formative years of human programming would have been vastly different. Yet, I would still be a spiritual being dealing with the human experience.

From my mind (and its assortment of Apostolic, Full Gospel Baptist, Pentecostal, Roman Catholic, Judaic, Islamic, and Christian religious influence), an overriding innate spiritual relationship is born that desires to add value to the lives of others.

I write this with the foundation of why I wrote this book, and that is to learn and teach the creation of neural pathways that connect to the positive energies in the atoms that come from the Breath of God, The Creator of All Heaven and Earth.

My desire is to create and manifest the original Atom of "Let there be light, and there was light,"/ Big Bang Theory. We now know that the original start of existence begins with the original energy force, science calls it an Atom. I am calling it the Breath that God breathed into the nostrils of Adam.

Things may not add up for your mentality the same, but for me, this works ideally, consistently pointing me towards positive perspectives in life, versus a negative outlook on life.

Most of us have heard the words, "born in sin, shaped in iniquity." Is it wrong to consider that man was born into a negative state due to Adam's sin, which then shaped a mindset and taught a plethora of "bad things"? Iniquity.

This is mostly done by other human beings who were also born into a negative state of sin, repeating the same cycle continuously.

We have no choice in coming into this world, but we do choose how we live in this world.

Is it hypocritical for an adult to desire to control another adult's free will?

If you would not want your free will to choose to be controlled, then it would be hypocritical for you to want to control another person's free will.

As a human being seeking greater answers to what is happening in the human thought process, particularly my own, I am overwhelmed with gratitude that I desire with my free will to be "wrapped up, and tied up," with my visualized words, labels, and phrases, God the Creator of all Heaven and Earth.

This means that I will always be trying to develop some way to compel humanity to understand the innate Power of Love, which lies within.

The power we know about but have not yet figured out how to master or control its effects on the human psyche.

The best explanation is the Power of Love and the Energy Force, which is the energy source.

God is Love and everything positive flows from that Love.

This is the journey of the spiritual being trying to master the human experience and the transition from what I used to think and do, to what I think and do now.

It was common in my thought process to remember the practice of not giving away all your secrets because, more often than not, someone would take advantage of you. To add insult to injury, the act would be followed by a statement like, "If he was stupid enough to do that," (put yourself in a position to be taken advantage of), "then he deserved to be taken advantage of."

This could be likened to asserting that you deserve to be robbed if you accidentally leave your car door unlocked at night.

This is a ridiculous and negative perspective to have programmed in your mind during your formative years, as a way to fill a void in the desires for an increase in capital. #dollardollarbillya'll.

All of these types of programs come from the subconscious belief and acknowledgment that slavery was so undoubtedly unfair regarding the brutal pay for the labor ratio, and most black Americans are subconsciously pissed at that fact, of the lost opportunity (of increased finances) that should have been exchanged for the energy of their ancestors' labor.

With the human desire for the reward that brings other rewards, we are left only with the solutions that we believe are available. At some point, we will subconsciously tell ourselves that reparations are owed and that it is acceptable to rob the system and its inhabitants.

We tell ourselves because of the formative years' teachings that the only way a Black person will ever possess forty acres and a mule in America is to take a shortcut to get there.

TRAS-R Thalamus Reticular Activating System Region

The subconscious twist that will manifest the thoughts of how to do just that... take a short cut.

My friend, my friend. The system and its inhabitants know human behavior: the likelihood of a person trying to take shortcuts to the reward because of limited opportunities.

There is an institution of correctional facilities that are aligned for the predicted influx of urban youth, who will not know how to create a neural pathway, which can shut down "The Hate U Give (THUG)."

This neural pathway thickens with the knowledge of "how to get my lick in," an example of that being the neural pathways created in my formative programming, growing up in Brooklyn.

Another strong example-- why do we learn what disrespect is before learning what respect is?

For instance, if you're on a crowded subway and someone steps on your toe, or bumps into you, and fails to say "Sorry" fast enough, it can trigger an emotional response that you are being disrespected as a person, and that it is very necessary to let society know of your displeasure.

Because of my programming, I would be looking out for who was disrespecting me, a lot more than thinking about who I was disrespecting.

This applied to everyone except my mother, who I loved immensely and didn't want to make her sad. I started out doing what she said, and then it turned into doing what she said and how much could get away with without her knowing about my activities.

This was the beginning of a conscious reprogramming of my belief system, growing from what my parents taught me to what I was teaching myself, and to what society was teaching me. Stimuli formed outside of my mother's control began to awaken my senses from slumber to a state of heightened awareness about how to achieve the rewards of money and power.

In hindsight, I'd consider my reason as the reason why I have made a lot of mistakes in life, and why things didn't work out well. My why came from visualized thoughts of the future benefits of having money in my pocket.

Subconscious behavior being brought to a conscious conversation

Because of the visualized thoughts we retain for our future selves, we will allow our current selves to ignore lawful instructions and take chances.

A negative neural pathway will make one ask: "What have I got to lose if I see my future self in a more negative way than my current self?"

The emotions that come with the thought of missing out are fully awake. This is because the stimulus that is familiar to me (the fear of missing out) was programmed and taught to me during my formative years to control my free will of choice. I did it for the cookie.

TRAS-R *Thalamus Reticular Activating System Region*

The Thalamus Reticular Activating System Region has a portion of its filter that will allow the 'Fear Of Missing Out' stimuli to pass through and have a lot of weight in the decision-making process.

If my situation is in dire straits, it is partly due to my upbringing in the urban city conditions of taught behavior, with those conditions becoming my belief system.

Until another neural pathway is created that overpowers the current pathway enough to take its place in my belief system, my belief system will stay the same.

Subconscious behavior being brought to a conscious conversation

Imagine you are behind the wheel, driving alone in your car--- your eyes are seeking, your ears are listening, and your hands are feeling as your sensory inputs are sending stimuli into your conscious mind. They are allowed to pass through the TRAS-R.

Unexpectedly, a car moves into your lane, and without conscious thought, a stimulus signal is sent through your neural network to the muscles in your leg, attached to your foot, to press extremely hard and abruptly fast on the brakes. SLAM!

You don't even think about your foot pressing the pedal, but the subconscious thoughts from the signals coming from the sensory organs bypass your conscious mind and move the members of your body. This is all without giving it a conscious thought like, "I am going to press this brake at this time, so this car doesn't hit me."

As we realize that in the time it took to read these words, consciously thinking about it would have been too late, and an accident would have happened. This is a good example of bringing subconscious behavior to a conscious conversation: showing a circumstance of your subconscious mind taking control of the members of your body to prevent harm. Survival.

Staying with the same theme, the moment when we first thought of driving a car can be described as the initial creation of our neural pathway, or nerves, in relation to driving. We created a 'folder,' so to speak, in the brain called "driving a car."

Now envision the things that belong in that folder, and the things that do not, such as 'things to do to help pass a driving test.'

We want it to become automatic, so we practice driving. It is during the practice of driving that we create the filters of, "yes do this," and, "no don't do that."

After we become proficient at driving, the things we first programmed "Yes" become automatic for the most part.

The things we said we should not do also become automatic, as in, "Automatically not allowed to pass through the TRAS -R", as something we should not be doing while driving.

First, focus, and make a conscious thought to stay between these two lines to avoid an accident. Next, focus with the anticipation that if someone comes within those lines, I already know that pressing the brakes will cause me to slow down and avoid the object that crossed the lines of my focus.

TRAS-R Thalamus Reticular Activating System Region

The Thalamus Reticular Activating System Region has done its job by sifting through the data and presenting only the pieces that are important to you. In this case, to press on the brake and avoid injury to the body.

This is an example of subconscious behavior being brought to the thought process of our brains. The questions become, "Should this be taught to a fourth or fifth grader?" And "Why is this portion of our education being left out?" It is a subconscious prioritizing of which stimuli enter the brain for your attention and awareness.

The top-secret deal is that we are responsible, along with the people from our past, for programming the TRAS-R to conceive the types of filters it creates or even knows it can create.

The programming comes from the stories we have told ourselves about everything we have seen or heard, touched, tasted, and smelled.

It is the reason you learn a new word and then start hearing it everywhere. Your TRAS-R takes what you focus on and creates a filter for it. It is why you can tune out a crowded room of people talking, yet sometimes hear something that sounds like your name, and you will detect that out of nowhere, even if you are in the middle of listening to another conversation. This is because of the filter that has been created since you were first called that name as a baby.

Any learning requires at least a minimal level of arousal, stimulation, and 'waking up' to pay attention to it, be interested, and concentrate on it. It is my hope that this curriculum gives that arousal and stimulation to your spiritual nature, and that the Spirit Man inside of you is led to create new neural pathways, which carry stimuli of GBGMGB (God's Brain, God's Mind, for God's Body), and that it would be allowed to pass through your RAS to your Thalamus, and the higher levels of the brain processes.

Lessons From the School of Hard Knocks, the curriculum, is the information that can and will be recalled for possible manifestation into a response by your future self. So, for our future selves' comfort, this seems like a very good place to reveal "The Why." as in "Why are we making the choices we are making for our future selves?"

The "why" is because we love ourselves, and that is a constant. The curriculum of LFTSOHKS is for the variable of emotions attached to our manifestations of the members of our body, which are the result of dealing with the issues of life.

The variables from the issues of life become the "how" we love ourselves and show love to others. How God loves us is the constant, and how we visualize our love for others is the variable. The variable will be based on the story you tell yourself.

This is the thought process inherited straight out Adam's original playbook, by visualizing the benefits for his future self with Eve, "bone of my bone, flesh of my flesh."

The act of choosing oneself is inherited from Adam's original neural pathway, which is now and is the second thickest (Love being thickest) neural pathway that's allowed to pass through the TRAS-R.

The Thalamus Reticular Activating System Region will allow in the stimuli that look for the benefits of choosing oneself, and the visualized benefits of a reward for your future self.

When we tell the story, what we tell ourselves will become the reason we do what we do, the why of our actions.

If we tell ourselves a story with hate or negativity attached to it, that becomes our why.

If we tell ourselves a story with love attached to it, that becomes our why.

Manifestation of His Image and Likeness

The manifestation of how you are about to do something is a conclusion of the repetitive thoughts of your why: why you are doing it in the first place.

A subconscious thought can be a taught conscious thought, thought about so many times that it becomes subconscious. If what was taught produces hate, then it can be reversed-engineered and untaught.

The innate love we were born with should be what we are left to think with as the constant of why we do, what we do, and how we do.

The subconscious spiritual question is right here.

Would it be a positive thing for my future self to align my thoughts, actions, habits, and character to the destiny of Oneness?

I align my thoughts, actions, habits, and character to the destiny of Oneness, and if do would it also be a way to prevent a decrease for my future self?

There are two stimulants that the stimuli that will be introduced to and your Thalamus' reticular activating region will be on the lookout for, first and foremost.

Put simply, *"What's in it for me, is it good or is it bad? If it is bad, how can I prevent it from happening? If it is good, how can I multiply it and increase it? How can I replenish it and use it to subdue others?"*

I know God meant to subdue all the other living animals of the Earth, but we have come too far mentally to ignore that mankind and his selfish greed for more has flipped this into subduing human beings as well. #Drop the mic. (spiritmanspeaks.com)

The new information says, "Let's bring up for conscious conversation the subconscious behavior of acting like we still have Sovereign Authority, as if we were still mentally One with God."

This is a place of mental manipulation, the place where the "twisted gets twisted" regarding our free will, who we choose to love, and how we choose to love them. This is especially true when we think of God, the Creator of all Heaven and Earth, who no man nor human has seen.

The neural pathway that connected Adam's mind to God was One. The original Energy Source of Love is the constant of why and how we love others, just as it is in the way we love ourselves.

Visualize a flashing inner light. The neural connection to love thy neighbor as thyself should pop up somewhere, and as you were visualizing those words, the synaptic nerve endings connected in milliseconds to your memory recall.

Now that we have this information, free will steps in to pass it down through the brainstem and nervous system and the motor skills of your body, determining what your hands do, where your feet go, what your eyes see, what your ears hear, and what your mouth says, that cuts sharper than a two-edged sword.

Why? Because words, labels, and phrases make the human visualize their meaning and subconsciously recall *emotions* that have weight in your decision-making process.

Adam's mental decision was a "less than one love for God" because he decided to choose love of self.

Now I have two neural pathways, the Love of God and love of self; conscious love for oneself while subconsciously experiencing the innate Love of God.

The constant of why God loves us comes from the innate breath that came from God.

The variable of how we perceive God love us comes from the taught stuff-- from our parents, teachers, neighborhoods, communities, and other influencers we have encountered.

Understanding the why and how variables will give you a much better opportunity to figure out what you are about to do next. because there is always a "What's next?"

Now we can create neural pathway blockers that shut down the pathway if what you are "fixin'" to do is known to be damaging in a detrimental way to your future self.

(Stand up and walk around with this thought for a minute.)

When we don't know what is going to happen next, it causes confusion, and allows the antagonist of this book to be the author of confusion. It is better to know what will happen than to not. Knowing the why and how will give a certain propensity to what will happen next based on the programming of your Thalamus reticular activating system region

The TRAS-R and Confirmation Bias

It is plain to see the tendency of thought of a person who primarily watches FOX News compared to a person who watches a network of a different perspective like CNBC News. We need to look no further than the politics in this country. So much so that there are rules that state politics should not be brought up or discussed at most workplaces. This is primarily because of its divisiveness, and often, the employer's knowledge that those types of conversations can disrupt employee behavior in the workplace.

But here's the trick--- by not having those conversations, both sides remain in the cocoon that was created in the formative years, never to come out to be the butterfly that affects positivity—the butterfly effect.

No new information is shared, no new dialogue is generated, and no new neural pathways are created that can challenge the formative years' thought process or belief system. There will never be a compromise of thoughts, and there will never be a way to understand each other's point of view. Even at the very minimum, this would still create a level of agreeing to disagree, versus thinking the other person is absolutely crazy for thinking of a conclusion that's so much different than yours.

LFTSOHKS, the curriculum, and the TRAS-R will challenge your belief system, acting as a designed tool for releasing the old way of thinking, and adopting a new way of thinking by creating new neural pathways that point to His Likeness and His Image. LOVE the way that God loves you, and through the TRAS-R, this is possible.

We look for things that we agree with and have chosen with our free will to create a filter in our brains to make our lives less complicated in being who we think we should be.

You could say that what we allow through the TRAS-R filter are the things on which we really want to focus. As we get older, we do things without much conscious thought because we have created such ruts in the pathways of our gray matter that it becomes an automatic link to our subconscious thought process; the repetition of the same thought, the same way, over and over that was created since our teachings, from the formative years.

The popular saying, "you can't teach an old dog new tricks," refers to the fact that someone over a particular age is highly likely to be set in their ways and unable to be taught or learn new tricks.

From a neuroscience point of view, this statement is misleading, but if someone continues to say it repeatedly, it becomes passed down as being a true statement. It will become a social norm, and society will stop trying.

Eventually, the older person will too, stop trying, settling with being "set in their ways", the way it was taught, and now the way it is believed.

This would be an example of how left-out facts and false information can limit a person's thinking due to the lack of mental stimulation, and visualization that can occur. In the human brain, when words, labels, and phrases are heard in the ear, the next thing that takes place is the transference to an electrical stimulus.

Stimuli have to move through the TRAS -R for you to pay attention to them.

TRAS-R Thalamus Reticular Activating System Region

Once you pay attention to a stimulus, the brain will give you five more seconds to convince it to continue to give this stimulus more "brain energy" to develop, and send the alert signal up through the brainstem to the Thalamus.

Have you ever started doing something that you would consider moderately important, only to start doing something else? The neuroscientific explanation would be that you gave more brain energy to a stimulus that passed through the TRAS-R, then continued on, deeming it important to your self-programmed desires, truths, and beliefs, and bypassed what you were originally doing. For example, "I am supposed to do my homework, which is very important, but I like and desire to watch television." For most of us, it would be hard to do homework efficiently with the TV on, as it would clearly be a distraction. This reason is why the TRAS-R will direct the brain energy for the attention of the TV program (because that is what *you* said was more important). This is the stimulus of all the "good" stuff, and virtualized benefits of watching TV, versus the stimuli that say you should do your homework.

The TRAS-R controls the ability to focus and sift through incoming information needed to manage the transitions between levels of the conscious mind and the subconscious mind and do your best. This controls what we prioritize in our current belief system.

Now, fast forward to report card time, and you get bad grades because you did not do your homework, which in turn affects your preparedness and productivity, causing you to get more bad grades. Now you get home and show mom, only to get your donkey-parts whipped, and the pain of that whipping is recalled from memory the next time you sit down to do your homework.

The stimulus for watching TV still gets through the TRAS-R (because you still like TV), but it is not allowed to continue because it is blocked by the stimuli that represent your recollection from the pain of your whipping.

Most humans imagine positive outcomes when we visualize things in our future. In this case, the positive outcome is less pain and better grades. This is exactly the lesson our parents wanted us to get, and we got it, some more than others. (Shout out to the older siblings who got it first, which allowed me to mentally record and learn from their mistakes.)

In the Thalamus Reticular Activating System, a filter is created to allow in information and the electrical stimuli that will help the body succeed in something that we (a conscious-subconscious deal) think we should succeed in.

"Better grades equal better education, equaling a good-paying job," is the neural stimulus that is created. This electrical impulse goes back and forth, all day, every day. As kids, our parents would often say, "It is your job to go to school and get good grades, along with a good education so that you can get a good job and get the heck out of my house!" And although this is all well, consider another perspective.

Subconscious behavior being brought to a conscious conversation. If this is the ceiling that most kids have been taught in their formative years, this will be the ceiling that they will aspire to reach, as there were not any pathways developed for any goals set above that. This limits the scope of what ideas can even be thought of.

Something needs to be done to take the limits off your brain. LFTSOHKS, the curriculum, contains "the why, the how, and the what" regarding God the Creator of all Heaven and Earth, and His design for the body of Adam. In short, the curriculum is the utilization of God's brain for God's mind, for God's body, and the manifestation of Love in the physical body.

The TRAS-R learns to identify patterns of encoded stimuli in the sensory signal transmitted through, and the more this cycle occurs, it gets a chance to make a bigger "groove" in the transmission line, which increases the speed at which the stimuli and information can travel (eight-lane highway versus two-lane highway).

For the most part, the key relation here is understanding that there is a direct correlation between setting goals and the TRAS-R When we look back at where He has brought us from, we can conclude that "ninety-nine and a half just won't do," and add an, "I guess I will run on and see what the end is going to be." - (*Lewis Jr., Rev. Howard. I Believe I'll Run on and See What the End Is Going to Be.*)

LFTSOHKS is the transitional workbook for putting off the old man and putting forward the new man--- mentally and figuratively speaking, of course. A new attitude will reflect something changing in your belief system. You will focus on that change, giving more brain energy to manifest that mental stimulus into your actions for the world and people on the outside world to see. The journey is about making it a great attitude and a loving attitude.

LFTSOHKS was designed to help with keeping it moving, away from the society-taught "elephant dung" programmed into us in our formative years. The TRAS-R adjusts the filters to your belief system as you program them to. The TRAS-R can now block and shut down the ones that don't make any sense to your peace of mind and stress-free living.

In most instances, things that smell and feel like crap tend to have society's crap attached to it. That feeling that people may say is a gut feeling, but it is actually sometimes a biological feeling of familiarity because of the electrical stimulus that the bull crap comes with. Until we engage our free will to accept different information that can change that, the TRAS-R will not allow that type of stimuli for further processing.

The TRAS-R can regulate emotions, which often feed into the weighted behavioral decision. The question becomes, "How much anger should I put in this decision I am getting ready to make? Or should I even put anger in?" Understanding how sufficient levels of cortical arousal and the connection to the other parts and members of the body work to inhibit impulses, and control strong emotions subconsciously could prevent one from having to ask themselves that question.

This is merely understanding how to stop it before it starts, by recognizing if you continue on in a certain type of thought process, you are guaranteed to go down a "rabbit-hole." The logical question then becomes something parallel to, "Why would I want to poke myself in the eye with a sharpened pencil? The predictable outcome of those actions clearly would not be beneficial to me." We can use that same mental equation in deciding how much value to put on recalled memories and the visualized projections of the expected outcome, avoiding the "rabbit hole" of negative emotions altogether.

Could it be the expectations of the Spirit Man side of the fence, or the expectations of the Natural Man side of the fence? What is the susceptibility of the brain functions to accept one side before the other?

Understanding how the mental agreement (between the subconscious mind and conscious mind) operates can cause a manifestation of action in the members of the body.

The intellectual ability to dismiss society- taught garbage by shutting down programmed words of negativity is probably most challenged when those words are used to purposely make a person of color feel less than.

Once I realized that in this sort of circumstance, words were meant to mentally limit my brain stimulation and allow the information that I received from the outer world to hinder my interactions tremendously in social encounters;

Once I realized that I was programmed from my formative years to believe the words, labels, and phrases, that were mostly negative, associated with being a darker pigmented person of the '60s;

Once I realized that this significantly affected the input into my RAS, preventing me from becoming my most authentic self;

Once I realized ALL of that, it significantly affected the stories I was telling myself. No longer could I consider myself less than; I now had every reason to become my most authentic self. #Drop the mic (spiritmanspeaks.com)

The information that you allow to pass through the TRAS-R should be for the benefit of your future self. Encoded stimuli in the sensory signals are transmitted through, and the more times this process occurs, the greater the chance becomes for a "groove" to be made in the transmission line, increasing the speed of the stimuli and how quickly information can travel (eight-lane highway vs. two-lane highway).

I figured out for my future self that the expectations and foresight of the "why" (as in "Why I should tell myself a story that I know will bring back a negatively charged visualized emotion) and its manifested actions were not going to be beneficial for me to carry out, in regard to dealing with life's issues. This is an example of the benefits of subconscious behavior being brought to a conscious conversation.

The premise is how to have a conscious conversation with oneself, knowing that it will subconsciously inhibit impulses and help control strong emotions recalled from the factual events in our lives, and recalled from our memory.

These are the things that are unknown and the unrealized, constant behavior that occurs in the subconscious thought process of our brains.

It would seem that something so pertinent yet simple, would be taught as young as the third grade. It is essential that this is included in the education of our youth. It is imperative to have a curriculum developed through non-profit organizing to produce a better understanding and provide education on how to articulate the why and how.

Selected cells of the TRAS-R are aroused or alerted when signals are transmitted through the assimilation of sensory information gathered from all body parts and all five senses. Imagine understanding the 'what' in relation to sufficient levels of cortical arousal and the connection to the other parts and members of the body, working toward manifesting something.

The 'why' message and information from the eye gate are chemically converted into an electrical stimulus, sending impulses to the entrance of the TRAS-R, and alerting the brain to its level of importance. If the 'why' matches your belief systems, it will be allowed to flow through. The why is established, and the "how to proceed". is the journey. What can be determined by the 'how'—as in "how to proceed"

LFTSOHKS, the curriculum, shows the 'how' in laying the foundation to manifest a set of subconscious protocols and subconscious behaviors that can be assessed on a conscious level.

As we mature in life, many of us do eventually learn how to reset the remote that was programmed for us by societal norms.

Imagine having a conscious conversation with yourself, knowing that it will subconsciously inhibit impulses and help control strong emotions.

We then can begin to understand the limits that have been put on our thought process through third-party programming and how the effects of believing false information can manifest as negative issues in your life.

Your conscious mind (5%) begins to tell you that you can't afford or don't deserve something, while your subconscious mind (95%) says it's okay. But, if your conscious mind creates and develops a positive neural pathway, it could see the same thing and change the perspective for the better. Instead of "you can't afford it," it becomes, "you don't have enough money to buy it yet."

This produces mental stimulation for the 95% of the subconscious mind to develop a strategy to execute.

Growing up in a marginalized neighborhood meant that, in order to get a nice car, I was going to have to do something illegal. If you grew up in a black American family that experiences financial struggles, there is a very high probability that these disadvantages can be traced back to a peculiar time, antebellum, and the peculiar institution, slavery.

TRAS-R Thalamus Reticular Activating System Region

The words, societal norms, and accepted unacceptable behaviors of the time remain in the subconscious programming of White and Black America, as well as its economic transgressions. These words and actions formed the mental chains around the necks of our great, great, great Grandmothers and Grandfathers, reminding them of the rules to stay alive.

Descendants of Black ancestry undoubtedly carry the trauma of simply having pride as well—and psychological weight of the ultimate inhumane practice, both mental and physical, of someone who believed they were superior, demanding that you either accept whatever label was being forced, or face the possibility of being beat until you did. This had to be programmed to a degree, eventually becoming a part of the "societal norm programming," with the TRAS-R allowing it to pass through.

"Peculiar Institution" is a euphemistic term for America's negative mindset involving making a mass amount of money off of my forefathers.

The stimulus from the third-party programming of the threat of physical death or bodily harm, or repercussions of not following the instructions from another human being, forced the belief that if I don't comply, this physical body will die. As the mind calculates, it also instructs the body and its limbs on what to do and how to do it. Naturally, the human brain wants no involvement with dying or being killed.

The TRAS-R plays a crucial and vital role, as it is the seat of our 'fight or flight' mechanism in the survival makeup of the human being. Largely in the minds of Black Americans are the visualized stories handed down of the relative that was hanged from a tree during the era of strange fruit, many times for nothing more than the color of their skin. The Black experience in America has such significance due to the tremendous efforts put forth to program our TRAS-R.

I challenge you to question the concept of Black-on-Black crime in reference to neural pathways and societal norms. Societal norms demonstrate that Black life is as disposable as an empty tissue box—throw away the empty one, and grab another. This subconscious rationale still prominently exists today, and in the famous words of a great lyricist named Kendrick, "So why did I weep when Trayvon Martin was in the street? When gang-banging make me kill a n***a blacker than me? Hypocrite!" *(Lamar, K., 2015- The Blacker the Berry)*

This is a direct reference to gang programming in our youth and the programming of being committed to getting the reward by any means necessary, believing that not getting it could mean life or death.

With that mindset, one is now mentally capable of taking a Black life because believing this nonsense results in the notion that Black life is worth very little and holds minimal value, thus leading to a mental mission segueing to a reward, money, or power—or all three. The idiotic notion that Black life is worthless, paired with the fact that a neural pathway has never been created in society to say otherwise, or that a Black life is as valuable as a White life, largely attributes to the confusion.

TRAS-R Thalamus Reticular Activating System Region

If no pre-existing neural pathway teaches me that Black lives matter and have value and worth, then these things won't be recalled when I choose to take the shortcut to money or power.

It is impossible to dismiss the fact that the White experience compared to the Black American experience has tremendously molded the TRAS-R of both sides. The Black TRAS-R has been handed-down, literally and figuratively, all the way back to the only neural pathway known, or the thicker and more well-traveled pathway that says a Black life does not matter or matters less—particularly if a reward is at hand.

Black people have a different responsibility in balancing out the thought process, and the TRAS altogether. Many times, the only recourse is to seek the help of our Higher Power. The TRAS has control of the body reflex mechanisms between subconscious thought and physical manifestation. It will become that much harder to overcome the unavoidable emotion of disrespect when the topic of slavery is brought up; and the sting of death is fueled by how much you allow societal norms to dictate your innate thought process, producing the 'what' for the rest of the world to see.

With the words, labels, and phases handed down being in place and doing what they were intended to do, neural pathways are built that will limit a Black person's thinking. Then you must choose to either continue to strengthen the existing negative pathways that make you believe you are undeserving; or recognize the reality of your struggle due to pre-existing socioeconomic conditions, and the 40 acres and mule that the government promised your people, but never gave.

The TRAS-R filter will only be looking out for information that fits inside our master program. This is the main story that we tell ourselves about who we are, what we represent, what we are willing to do, and what we would never try. These things then manifest into actions as we deal with the issues of life, reinforcing and thickening the created neural network.

It does this by calling our attention to any pertinent information, which would be considered background noise otherwise. The loud bang of something dropping on the floor while you were reading your book-- the loud noise did not have anything to do with what you were focusing on. Still, the noises that you collected from your ear sensory made it through the filter, versus other outside noise, because you were already programmed to pay attention to certain information, or stimuli that carry the encoded information. These are the types of bits of information that immediately become more important than the book you may be reading at the time.

When brain cells communicate frequently, the connection between them strengthens, and the messages that travel the same pathways in the brain over and over again begin to transmit faster and faster. These behaviors become automatic with enough repetition, like reading, driving, riding a bike, and other non-complicated behaviors that we perform instinctively because of formed neural pathways.

I invite you to look at every video you can find online that explains the Reticular Activating System from a neuroscience point of view. I also encourage you to look at videos that explain the Law of Attraction, as well as the practice of visualization.

TRAS-R Thalamus Reticular Activating System Region

Anything that you have ever heard from a motivational speaker, or life coach of some sort, which made you feel super excited about reaching and achieving new heights in life, gave you that sensation because of a large volume of electrically charged dopamine and serotonin being released into the brain.

Visualize your health improvement, beginning with something like going to the gym. Start by visualizing the joints and the muscles in your body, staying fit so that you can healthily continue on in your desire to share as much Love as you can. You're going to see yourself taking care of yourself, and when you start to spiritually visualize The Image Whose Likeness you were made in, things will happen. When your eyes are open, you tend to look at other sights and get distracted. When you close them, one of your senses isn't being used, leading to less distraction.

Learning new ways to visualize is beneficial because, when you encode, what you encode becomes a real memory. These memories can change the thought process, giving you the "seeing it before you see it," feeling, or déjà vu.

Scientific research shows that visualizing aids the manifestation of creating, developing, and improving the skills just as if you were physically performing them. Visualization can produce the desired results, with its connections lying deep within the brainstem, between the spinal cord and the brain. As the number of dendrites increases with the frequency at which a behavior is repeated, a thicker neural pathway is developed in our brain. When brain cells communicate frequently, their connection strengthens, and messages repeatedly travel the same pathways to transmit faster.

God's design of the TRAS -R is where the dynamics of choice are chosen. God designed this filter to assist you in sorting out and doing away with the things that are not like Him. So much of our time is spent thinking negative thoughts that it seems we have the ability to talk ourselves out of something in roughly five seconds.

It's similar to when we ride through an extravagant suburb or a newly built neighborhood— you see the beautiful houses and begin to fantasize, but your flesh-governed brain quickly talks you out of it. You begin to dissect the reality of your own finances, and the final thought becomes, "These homes are unachievable, for me. I don't know how, in my mind or body, to make something like that happen for myself." Your mind will stop looking for ways to make it possible, and your mind will stop seeking out people that can help you. Your mind will stop trying to learn how to make it attainable and limit itself to what you believe there is no need to pursue. After reading this book, you will clearly see that it is time to create new neural pathways composed of genetic material, positivity, and hope.

You will understand 'God's brain, for God's mind, for God's body,' and how prayer is the creation of genetic material.

Our DNA originates in prayer. By focusing on the thought of that particular type of materialization of a desire, the DNA of the material (that of which you desire) and the functions of the brain and mind will manifest the DNA to create that prayer or request in your life.

Praying is for manifesting things into your life. It is a request from your conscious mind to your subconscious mind to make something happen and figure things out. The subconscious mind then pulls things together in mathematical calculations and comparisons that form an equation, and a logical conclusion. You then manifest the results into your actions to make it happen, whatever it is that is happening in your life.

So now you make the initial request, and the next thing is how much do you focus on the request? This equates to the amount of time and energy it takes until you have manifested your request. The number of times you return to that initial thought as you focus on your said desire will affect your process of manifesting the actions of your body to create what you are focusing on.

You have no control over what people may say or do, but you do have control over how you respond. This is called E+R=O, or Event + Reaction = Outcome.

The Word says to "Lean not to your own understanding," at the occurrence of an event, and "acknowledge God, and He shall direct your path" -- reaction and outcome. The thought of God's Love towards me encourages my Spirit Man to trust wisdom and walk into my foresight.

Calamity is your brain trying to figure things out without God's word. After doing the work taught in LFTSOHKS, when you attend church, you will be able to multiply the meaning in the messages and manifest them for your purpose, for which God designed you. However, we become more hyperaware, and the ability to sense when people may be doing us wrong is sharpened.

Confirmation bias, or 'myside bias,' is the tendency for one to seek, endorse, or recall information that directly supports or confirms their perspective. When this occurs, individuals who hold a strong position search desperately for information that confirms a previously held view.

Confirmation bias is important because it can lead people to hold strongly to false beliefs or give more weight to information that supports their beliefs than what is warranted by the evidence available. In cognitive science and psychology, confirmation bias is the tendency to interpret or seek out information in a method that will confirm one's preconceptions, often leading to statistical errors.

> **"Wisdom is the principal thing; therefore, get wisdom: and with all thy getting get understanding."**

With this knowledge, I am able to get a better understanding of any opposing party's decision-making process, and work to form understanding from their perspective. I can consider the situation or problem, and not the individuals, leading with empathy, in all areas of response based on Love.

Putting myself in the other person's shoes could give me better clarity on how to add positivity to a tense situation. It is important to be knowledgeable of these biases, which can influence and mold my perspectives and decisions.

TRAS-R Thalamus Reticular Activating System Region

God Brain God's Mind for God's Body

The conscious mindset of your thought process is produced from your belief systems. We look for information to put back into that belief system and allow certain information to pass through the TRAS-R because we want to confirm what we believe.

GBGMGB will help you uncover the biases, beliefs, and perspectives that impact your decisions and allow you to determine if they are pointing towards God and Love.

Listing the ways that your unconscious biases impact your thoughts and behaviors can assist you in starting to make better decisions and alter your thought patterns by shutting down the old neural pathways that aren't going in the right direction and creating new neural pathways that do. Again, we look for things that we agree with because it reinforces the neural pathways that we already have. New can mean a different way

Thankfully we have the scientific knowledge of neural plasticity and its functions.

I'm not crazy, and I am not trying to start a new religion. But I am in LOVE. I love the Lord with all my heart and soul, and I want you to have what I have. This book is for those who will insert the Word of God into our minds to grow in new filters that are designed by God. It is time that we do the work for God's brain so that the 'why' (we are made in His Image and Likeness) becomes bigger than what society has shown and taught us.

God's brain is fully capable of removing old information while filtering through new information. New information will encode into the TRAS-R in the brain and body that God made for His Glory and Purpose, there is a stem-shaped portion of the brain controls most of the body's involuntary functions and reflexes-- what your hands do, where your feet go, and what your big mouth says. Man has labeled this stem-shaped portion of the brain The Brain Stem. Within the Brain Stem is the Thalamus and Reticular Activating System. When the belief system of a human being changes, the new ideals are taught through the Thalamus Reticular Activating System Region or TRAS-R.

My TRAS-R was set up like a heat-seeking missile on the lookout for money-making opportunities. The things and opportunities that I could, or potentially, get away with, are now programmed to be allowed through my TRAS-R, immediately changing my thought process to the surroundings of my socio-economically disadvantaged, welfare-receiving neighborhood.

I think most Black people over the age of fifty can agree that welfare, in a way, taught us how to game the system, but the real trick became the reality that the system was gaming us. The plot became twisted when the premise became limiting your drive for real money by thinking you were getting easy money.

While on the subject of "easy money," once the basic human needs of the mind are met, it doesn't put more energy towards accomplishing that task or need, until it is needed again. Considering Maslow's Hierarchy of Needs, the drive dissipates with food and shelter being provided.

TRAS-R Thalamus Reticular Activating System Region

Here is the trick. Let's set the scenario--- the government and the human psyche meet at neuroscience plaza to have a conversation: The government says, "I will give you this free aid, and in turn, you will use your brain energy to complete the other tasks that make you a good, tax-paying citizen."

The human psyche responds, "Good! I will take the reward of money and power."

Again, neuroscience shows that the human brain prefers to take the path of least resistance. So, if you give me shelter and food, I have no real mental stimulation or urgency in obtaining future needs. Because of recalling and visualizing, it is at that time that dopamine and serotonin are released, satisfying any stress triggers associated with food and shelter, or anticipating the "1st and 15th."

I am proud to say that my mother figured this out after being on welfare for seventeen years. She returned to college at 37 and received her bachelor's degree, furthering her position as an outstanding, tax-paying citizen.

Who is at fault in this vicious cycle of dependency? Should welfare include more fishing classes, in their programs, adopting the logic of teaching a man to fish who in turn, could feed himself and his family; as opposed to giving him fish, and diluting his drive to create, resulting in remaining marginalized?

Because my mental aspirations were limited to constantly preparing to get the reward of money and power, I started hanging out with a little gang called "The Bad Ones." At a young age, this was mental reinforcement, creating a neural network that said it was 'okay to be a bad one.' I was definitely in the bad boy category and 'going along with the gang' for the teachings of my formative years. This indicated that there was power and safety in numbers in regard to being preyed on. More importantly, it showed that there is power in the number of those you could prey on to take shortcuts to the rewards of money and power.

I bring these formative years of programming to the forefront for the purpose of conscious conversation. This is the foundation that a large amount of urban youth must overcome, or better yet, be taught. It is the notion that these types of created filters lead to the set up for jail and prison, chasing after the money and power with limited information, and limited brain information.

The saying goes, "If you know better, you do better." Eventually, you learn your lesson on how not to manifest certain things in the members of your body; because now you have physical facts proving that if you repeat the same manifestation of thoughts, the outcome will probably not be a positive experience. The flip side of that coin is, if you don't know that you can do better, you will not do better, because of the lack of information on how to do better. #This is the type of lack of knowledge that causes people to perish.

Wisdom speaks

You probably have figured out by now that sometimes in life, it is better to slow your mind down a bit so that new information can come in. This is one of those times.

The Thalamus Reticular Activating System Region forms nerve cells and their connections lying deep within the structure, between the brain and the spinal cord. It activates the entire cerebral cortex with electrical impulses of energy, increasing the level of arousal and readiness (or waking it up) for interpretation of incoming information. It works to filter out unnecessary information so that the information deemed important for what's next gets through.

Lessons from the School of Hard Knocks— because there is always a what's next, the lesson is, what are you going to do about, what's next?

The path of figuring out how to go about doing it is a lot of times figured out based on life experiences. It is by recalling the experience and visualizing the experience and how it was done. The mistakes in our decision-making come from a lack of understanding of why we are about to do what we are about to do.

The goal of LFTSOHKS is to communicate to the readers of this book my description of where to focus with regards to our belief system.

The human minds that are programmed in human brains all over the world stand at the crossroads of free will and choice.

Two variables going in one resulting manifestation coming out, for the world to see.

Your conscious mind commands and your subconscious mind obeys. Your subconscious mind is an unquestioning servant. It works day and night to make your behavior fit a pattern consistent with your emotionalized thoughts, hopes, and desires.

As we deal with the issues of life, your subconscious mind grows through the flow of positive energy, and by understanding how to program God's brain for love.

The reason that divide-and-conquer tactics can and will continue to be successful in humanity is because of the many different variables that can entice the greed of a human, including the killing of his brother for an anticipated and expected reward. Think Cain and Abel—there is nothing new under the Sun.

The mind does not know the difference between when you are practicing something mentally and when it is truly happening. It produces the same emotion as if the event was actually taking place.

Imagine yourself giving an important presentation. Visualizing your presentation and practicing or *pretending* can help you knock it out of the park and remember all of your points.

The happiness you feel from this success triggers a dopamine release. As you create this graphical motion picture in your mind, you will take notice of the noises, dialogues, and particulars of that moment. As we repeat this movie in our heads, this helps to create a thick neural pathway that is allowed to pass through the TRAS-R

God's Brain God's Mind God's Body is about visualizing what we want, followed by allowing our unconscious and conscious minds to collaborate and make it happen.

GBGMGB helps the reader to understand the information, take it apart, and reverse-engineer the catalog of components that produce negativity and hate.

GBGMGB then puts it (our mental thought process) back together again. "You are what you believe yourself to be." It seems that this would be a very good reason to know the 'why.' as in why we are making the choices we are making for our future selves.

How God loves us is constant, and how we love others varies as a result of the things we tell ourselves about other people. However, we don't tell ourselves a story of the constant. We tell ourselves stories of the variable because of the selfish nature inherited from Adam. The variable is the doorway to confusion, allowing the antagonist of this book to be in opposition.

Why would one person want to display negativity towards another, and what are the mental effects of saying something meant to be derogatory? What is the benefit to the person who just made the insult?

It seems that there could only be one logical reason for a person to want to make another person visualize and recall a negative, and that's for mental manipulation, knowing how the attached words, labels, and phrases will affect the human psyche.

I learned later in life that this was what Black people often labeled and dismissed as just some "stuff white people do," using words as a way to harass the Black mentality and to get them to mentally visualize their ancestors hanging from a tree, creating the subservient mentality, and diminishing my ability to act on my own discretion because of the constraints of foreseen fate. Even today, this occurs under the umbrella of mental manipulation and systematically racist tactics. There is enough history of non-conforming to the inequalities that can now clearly be seen by another human being. In short, God is Love, and if you aren't representing God or Love, what is your thought process filled with?

LFTSOHKS will shine a great light on the negative producing behaviors of the subconscious mind, which do not produce the love we feel from God, and every day you visualize walking by faith and not by sight.

When you visualize yourself doing things that build new neural pathways back to God, you develop and improve the skills just as if you were actually doing it in the members of your body and mind.

If Adam's fault in sinning was his disobedience, then so was his mental conclusion of an issue, a circumstance, and entanglement. In all intellectual fact, there can only be one reward, the Spiritual Reward, and it can only be understood through visualization of a Heavenly reward to be had after this bodily death.

The power to succeed lies within you, and it is essential to learn how to tap into those powers and abilities within you to achieve this higher quality of life. Whatever name science might have attached to this ability of the brain to perform the way it does, does not change the fact that God designed it with qualities that would please Him. Are you using your brain to manifest God's thoughts into the physical members of your body? The reflection of your belief system of God's Love for you should be reflected and manifested in the members of your body for all to see. Do the members of your body reflect how you feel about God's Love for you?

It seems that the journey we are all on is about understanding how to make the conscious and subconscious thought process sync together in Oneness. GBGMGB will show you how to pair that desire with consistent actions. As you are in Him, He is in you.

God's design of the human brain for the body of Adam has to have a place for free will to do its business. It needs a place to decide, and a place where the rubber meets the road with what path you will take.

The choice lies between the path that will bring back negative emotions or the path that will bring back positive emotions. It is the difference between a path of confusion and delay, or clarity and forward movement—the space between stimulus and response.

If you believe this statement is true, you are able to rightly divide the Word of God when mentally stimulated.

When stimulated by the Dunamis of the Word of God, our response derives from a spiritual point of view. When stimulated from the outside world, the information retrieved has to go through a filter before it is allowed to go on into the Thalamus and other higher processes of the brain.

With the manifestation of the actions of the members of our body, the exchange between stimulus and response needs a place for this to happen-- a place where free will has freedom and power to choose what brings freedom, with the option to grow into our most authentic selves.

God's design in the brain dictates that this is the place where your mental activity is engaged in a series of changes before any conclusion is reached. The weight of the pluses or the minus, zeros, or ones, positive or negative, go or no-go logic is calculated here.

It is the space between a sensation to do and responding to that sensation with a logical, verbal articulation or physical manifestation. Understanding the programming of the TRAS-R would lead to a better understanding of what we are planning to do and why, because there is an understanding of where the story comes from, and the emotions that are attached to it.

The result is being in a better position to remind ourselves why visualizing the labeled issues that bring negative emotions into the thought process is not productive or healthy. When we bring subconscious behavior to a conscious conversation, we can better understand what subconscious behavior can do in terms of manifesting its own agenda to the members of our body.

TRAS-R Thalamus Reticular Activating System Region

The manifestation of our actions receives its guidelines from the Thalamus Reticular Activating System Region. This space we can call our free will attention span, where we get a stimulus from the outside world through our senses and choose to respond to it, or not.

That's free will in action. To understand that action, you have to understand the design by God that permits or denies entry to information that has a chance to become part of one's belief system.

More importantly, because of free will, we get to decide how our belief system is programmed and maintained. The TRAS-R is an area deserving of being highlighted because the train of thought begins after this filter.

When we try to imagine the mind of God, the best we can do is to use our understanding of how our own brains work that He designed. It only makes sense that the brain God designed for our minds would be incredibly sophisticated, with a navigation system built into the original intent of how to love, and how to choose love over hate.

This is where the struggle occurs in the "battlefield of the mind." The battlefield of visualized words, labels, and phrases that invoke chemical-changing equations in the brain and produce the slew of emotions that will become a part of your belief system, manifesting into the members of your body.

This designed portion of the brain is a filter that sorts through billions of bits of data at any given time, which must be in an organized order of importance to prevent a brain explosion from information overload.

Take the time to visualize an executive secretary that rightly divides the Word of God, filtering out unnecessary information and allowing only essential information to make it through. Only information important to the CEO or Mind of God is permissible.

Now, visualize a very large nightclub bouncer working security at the door, equipped with a list that determines whether or not you can enter, and the power to decide that your wardrobe isn't welcomed.

The same principles apply here, except the stimuli must be dressed a certain way to get in. The little-known part is that we are the designers of the dress code. We are the ones who inform the executive secretary of her duties,

If you love positivity, you will become more aware of and look for positivity in your surroundings. If you look for negative situations, you will become more aware of the negativity in your surroundings.

A sleeping parent can live on a busy street and drown out the noise of traffic. That same sleeping parent will hear their young child cry out in the middle of the night and immediately snap awake with concern for the child.

This is because the stimuli are dressed in the right things that have a high tune of alert. The stimuli that pass through signaling "my baby is crying" do not have the same stimuli as the noise from the street.

This is the same effect as wearing the right suit or dress for the lounge bouncer to let you in, or the executive secretary saying, "No, you cannot proceed to the higher intellect for processing because your information is not on a list of things already deemed to be important."

Thalamus Reticular Activating System -Region
(pronounced TRASER)
And Gods Brain for Gods Mind for movement in Gods Body

It makes sense that God's designed use of the TRAS-R and the information collected by our sensory inputs would allow data that supports positivity and spiritual growth.

When the TRAS-R is working and operating in its original intent, it is moving in favor of GBGMGB by establishing intentions that point towards Love. #His Image #His Likeness.

After intent is established (imagine the lesson plans from school being designed to teach intently for your mental processing), the brain works to give you a concluding answer.

We automatically perform many of these things, thinking about them through our subconscious mind. How many people have to think about what a letter of the alphabet looks like? How common is it to confuse A with B or C with D? We were taught how to recognize the differences in the alphabet the same way we can be taught to recognize the difference between love and hate.

We can now be taught how to analyze what is wrong with the conclusion, and when it does not look like God, or when it is not recognizable to the original intent of what was taught. In my opinion, the root cause can be traced to the story we tell ourselves, ranging from the society- taught hate to the existence of His innate Love—in His Likeness and in His Image.

It seems that it is imperative to have the right intent in order for things to work out in your favor. When you show love to another human being, you get a positive response back most of the time. When you show hate to another human being, you will receive hate in return, almost all of the time.

How did the wise man get wise? He programmed himself not to be foolish. Making a change and programming yourself "not to be foolish," would require you to determine which acts of repetition you need to use to shut down the old neural pathways leading to negativity, and visualize what is necessary to create new, positive ones.

If it was manifested in GBGMGB, then it is going to work in your favor. "For as a man thinketh in his heart, so is he."

This happens by taking the subconscious thoughts of our spiritual nature and marrying them to the conscious thoughts of our flesh nature. This is what occurs when the brain thinks, "I want to do the right thing, but can I do the right thing?"

I firmly believe that his psyche circles back to Adam's fallen thought process. We see and hear what we would like to see and hear.

In doing this, we are influenced by what we manifest into our actions, to see and hear what we want to see and hear.

Focusing on the negative things will leave you doubtful, with a 'glass-half-empty' outlook on life, versus focusing on positive things, and a glass-half-full outlook, which leads to hope and producing visualized images of increased manifestation and productivity.

Stimulus plus response equals manifested actions! What you think about, you bring about. It requires continuous improvement, and if you don't manage and maintain your thoughts, someone else will.

How do you renew your mind to choose differently from your past mistakes? As you choose differently, you will subsequently create new neural pathways.

The Lord will be with you wherever you go.

Visualization activates your belief by building up your faith whenever you hear and read the Word of God. If you grew up in a family or house where you frequently heard the Word of God being read and spoken, your thought process would likely understand how to recall that memory with the Word of God attached to it.

You would know how to visualize the Greater that is in you, and in my opinion, so much more of that is needed in the world.

On the other side of this, without this connection, your TRAS may have difficulty distinguishing real events from the made-up stories you are telling yourself, and is a root reason of why God is not very fond of dishonesty. In other words, the TRAS-R tends to believe whatever story you give it. to choose, the words, labels and phrases *you* allow into your thought process to be visualized and imagined for processing in the brain functions; all the way to manifestation of the visualized thoughts of what your hands might, do, where your feet may go, and what your mouth might say. #The Seat of Free Will

This is why in arguments and disagreements, the things said and the things that are received can get misconstrued, leaving you feeling perplexed or befuddled.

This is due to each person's TRAS-R filtering allowing some things in, while keeping a lot of things out. Especially the things that you don't agree with.

This explains the fact that, if applied properly, pre-marriage counseling can have God-intoned effects on the extremely high divorce rate.

The connection of dendrites increases with the frequency of use, and with repetition, the neural pathway gets thicker and stronger with habit.

Think Steph Curry's precise wrist and consistent execution. Repetition in learning new habits will eventually become that new behavior.

The rEVOLution is about the empowerment of love to accomplish what God has placed you here to do, your purpose-driven life. It is my intention to empower through the Spirit of God in man, to overthrow the flesh nature of any fallen man.

So, the question is: Can we mentally embrace to the extent that we can reprogram our minds with the Word of God that equals Love? We can do this by using the brain that God designed, for His Purpose and Glory.

By using scientific data, we can change the mental model of the way we understand God's Brain, for God's Mind, and for God's Body.

The use of repetition in this book is deliberate because repetition creates and nourishes new neural pathways that lead to memory. This is how you create a new neural pathway with the Truth; a neural pathway that is grounded in Truth, and a neural network, which can shut down old ways, knowing that the Power of the Truth leads to the Power to overcome everything.

We are to walk by faith, and not by sight. It is key not to look at your circumstances as just what meets the visual eye. From a neuroscience point of view and in reference to visualizing faith, we have knowledge of the benefit of increased focus due to visualizing.

Lessons From the School of Hardknocks

God, the Creator of all Heaven and Earth, designed your brain and your mind to visualize Him and His Love for us, increasing our focus to maintain the Power in His Love.

Chapter Seven: Seventh Grade

It's Easier This Way, The Path of Least Resistance

The path of least resistance is the least resistance it takes to move forward and the least resistance for positive increase by a given object or thought process among a set of alternative paths and alternative facts.

I could have experienced a number of benefits, as early as ten years old, if I had known how my "brain worked" regarding procrastination and the path of least resistance.

The way water flows around the rock, and not through it, is an example of a path of least resistance. The human brain has a foundational principle: to be as efficient as possible.

The brain does not have time to procrastinate unless you tell it what to procrastinate on. The brain does not have time to "delay." In fact, the importance of time is one of the primary reasons most people choose not to engage in garbage rhetoric and junk. Of course, the older you get, the more this foundational principle comes into your thought process, which again is visualized as the sand in an hourglass.

This is the same subconscious mental mechanism that goes into effect, with regards to being able to shut down all incoming information. If it is not what we want to hear, we have the ability to not hear it, and not process it for further thought.

Gas Mileage

If I am offered a chance to drive 45 miles every day to work versus driving 15 miles, the path of least resistance will come into play if we will take the job or not.} When we have a visual image of things and they appear far away, the path of least resistance will be engaged without us even realizing that we are killing a dream.

Visually, when our eyes see something far off, the brain has less desire for it. Take, for instance, desiring to go to McDonald's for food when there is a more healthy option less than one block away. Now factor in the nutritional meal being 15 blocks away, while McDonald's is only one. The subconscious equation becomes "Because it is easier to get McDonald's down the block than a healthy meal 15 blocks away, 'Time is money, and I'm hungry'."

We know in our minds and have visions of fresh vegetables on display, but because of the distance away, or the short amount of time it takes to get the fast food, the mental pathway of least resistance is stimulated. The least amount of physical effort is also added to the equation, bypassing the scientific facts of information relating to the good nutrition of fresh vegetables.

Now over time, and through the findings of your annual physicals, you start to realize the consequences of your choices --- choosing foods that promote a high-density lipoprotein (good cholesterol) level, versus foods that support low-density lipoprotein (bad cholesterol). The brain that God designed for the body does what it was designed to do by rewiring and reprogramming itself to include this new information.

God most certainly has a sense of humor. He really did make this thing simple, and despite the challenges, I feel good that we are all on this journey together in figuring out just how humorous man's folly can be.

When it comes to putting off the old man and putting on the new, you are changing your belief system from what was taught to you in your formative years.

Subconscious behavior being brought up for a conscious conversation.

It can be done with more clarity, specifically when you understand the scientific information that goes along with mental procrastination and the path of least resistance.

The path changes from the inclusion of "time is money, and I'm hungry," to the path of least resistance of not wanting health problems from the convenience of bad food choices. Newly created neuronal equations include this information, and the healthy options become a "no-brainer," because the brain's number one job is to protect the body from unnecessary harm.

However, these innate protocols can be bypassed and easily manipulated by the words, labels, and phrases created by some very smart doctors or someone of the sort, I'm sure; especially considering that they have figured out the human DNA codes encoded as the human genome and how they interact with each other.

If society norms, and perhaps an antagonist of the messages in this book, can get you to look at life as if you were small and insignificant, the opposing energy has done its job.

Again, there is nothing new under the Sun.

Some of us may recall from our memories the preachers and reverends of our youth telling the Bible story from Exodus, of Caleb and Joshua and the other ten men who were sent to scout the land of giants. There were the ten who saw themselves as grasshoppers; the ones who saw themselves as small and insignificant, believing that the other men were too big, too strong, and could not be defeated.

This created a path of least resistance and shut down the process of mental creativity and inspiration, removing the variables that composed their real problem of how to overcome, how to create forethought, and how to have strong faith and belief in that foresight.

The brain does not have time to procrastinate because it is busy doing all the other amazing things that it does, sustaining all bodily functions. Because of the designed efficiency built-in, the brain has a "problem" with things that it thinks may pose a problem.

When the brain identifies the labels as being hurtful, painful, or overwhelming, it will require extra mental effort to get past them and figure out a path of least resistance.

Consider Newton's' Third Law

"To every action, there is always an opposite and equal reaction." All forces occur in pairs, such that if one exerts a force on another object, then the second object exerts an equal and opposite reaction force on the first.

When we use certain words that we know will have an emotional effect, the equal and opposite force of the person wanting to use negative words and phrases to make someone feel bad or inferior would be that person having a subconscious desire to feel good or superior to another human being.

In the human brain, the scientific fact remains that when feelings and emotions are brought up subconsciously, we recall and visualize what these words mean, as they were taught to us initially, to mean something. "You can't be *this* because it is in my belief system that you can't."

I can only imagine the intellectual stories that Marlon Briscoe, a former quarterback for the Denver Broncos, had to tell himself before being the first starting Black quarterback for the team.

This leads me to recall my imagined guidance counselor training during the '50s, '60s, and '70s, the subconscious programmed behavior of the White guidance counselor doing the job the way it was taught to him, to do it.

But by whom, and based on what? Deceitfulness and deviousness in systematic maneuvering to manipulate the mental desires of another by opposing human beings and restricting the creative limits that open the door to raising one's consciousness. #Drop the mic (spiritmanspeaks.com)

Words, labels, and phrases of secret dishonesty. Words, labels, and phrases that could and will be recalled at a later date and put into our decision-making process.

Words, labels, and phrases that could, and will, restrict the growth of neural pathways into an awareness of what the individual's mind is capable of doing.

Guidance counselors were taught and believed that Black kids should not pursue certain careers because that was the information that they were given. Apparently, it made sense in their white, superiority-infused brains, further supporting systemic dishonesty.

It was taught to them a certain way, and the reasons behind this became a part of the White guidance counselors' belief system. It is very reminiscent of White realtors during that time using unfair and discriminatory rules to redline and deter Black homebuyers.

Procrastination

"I can accept the word lazy and its connotations, or I could get a better understanding of what is going on in my brain, with regards to what triggers procrastination."

By having the knowledge of what's going on in the human brain functions, we add value to the information we tell ourselves.

This, in turn, adds value to what you and I think and adds value to how we feel and what we believe.

We will have better information to draw from in our decision-making process. With these new values, I can make a better decision on "What's next?" because there is always a "What's next."

Even if you fall face-first into the concrete because you drank too much or took too many chemical-altering drugs, affecting your equilibrium, motor skills, and neural connections...there is always a "What's next."

It is better to be prepared for it, and that, my friend, is where the benefits of wisdom start to have more weight and value in our decision-making process. We should know how to insert that new level of consciousness to focus on that increased level of knowledge.

This is the formula to help to see more clearly, and more often, how to use your newly rewired brain. You are now adding wisdom to your decisions that you did not have before. You are changing the words, labels, and phrases that you tell yourself.

You are rewiring your thought process to add up to something else, and because of these newly encoded bits of electrochemistry, there is now enough power to kick your old neural pathway to the curb and out of your belief system! #Santa Claus

It is my hope that this chapter is articulated in a way that will make the reader visualize the words and add their value up to be mentally stimulating enough, and good enough, to pass through your TRAS-R.

If someone can get you to waste your time, then that really means you will have less time to stay focused on the things that matter the most. It's going to cost you something.

Subconscious behavior being brought to a conscious conversation.

Chapter Eight: Eighth Grade

Procrastination and Colored Peoples' Time

PROCRASTINATION HAS BEEN coupled with the negative implication of laziness. It is perceived as a negative trait due to its hampering effect on efficiency in reference to time and time management. So, when we go by the words, labels, and phrases that we were told and what they meant, our neural pathways begin to identify with the label of the pathway when it was created. That same neural pathway grows with repetition of the same electrical impulses to do something the same way, over and over. Think, "ABCs and 123's."

It is when we agree, with ourselves, and accept it in our belief system that we start to produce the same electro/chemical combination of our newly raised consciousness of our beliefs. It is in this newly created consciousness that we have new opportunities to create things out of that consciousness. We put one plus one together to get two, which created another level of consciousness in your ability to create consciousness. Many people reach a level of consciousness where they know the power of love because they are consciously aware of their experience(s). When we recall these experiences, we are often not conscious enough to control the emotions that come with the visualized experiences. This could consist of good and positive emotions and bad and negative emotions. Take, for example, the idea of society pulling the wool over on our physiological eyes. So much so that we have allowed societal norms to be embedded in our subconscious as a truth.

Lord knows I love all human beings, but these are lessons from the School of Hard Knocks. CP Procrastination is near the top of the list as a big killer of the dreams of Black people because, during the formative years, the concept of time and self were passed down as acceptable for us.

Time equals money and has to be factored into the equation at some point. To a larger degree, the perception of late becomes lazy, irresponsible, and unreliable. I use these words because if you take a minute to visualize these words, they will probably bring about associations of being late.

If you were starting to form a collection of human beings to do a project, would you want these qualities in the character of those you are delegating to complete this project or job? I only bring this subconscious thought process up for a conscious discussion. When the human mind (White or Black) hears words, labels, or phrases, it will visualize and plug into the neural pathway of what that word, label, or phrase means to them and their belief system.

We were told and shown, and then showed and repeatedly told ourselves that being late was something Black people do.

Circa 1950, Black people knew it was best to travel on back roads in some places, but it also might make you late. It was worth not having to deal with white authorities or stopping somewhere to eat, or sleep, only to find out that they did not lawfully have to accommodate you. Was this the origin of CP time?

Probably not, but the point that I bring about is that the human brain will take the path of least resistance, to avoid some things or a lot of things.

A Black man must work twice as hard as the white man, so the hard-working Black man believes this to be true. The path of least resistance to keeping a paycheck coming in and food on the dinner table was to do your work twice as hard to overcome the disadvantage of confirmation biases by white supervisors and upper management in charge. I have to work twice as hard to help the supervisor overcome what was taught to them in their formative years, knowing that when they see you, they are looking to satisfy their confirmation bias that suggests Black people are lazy when it comes to working.

If you were a slave, how fast would you jump to be back out there in the hot sun picking cotton for "almost" free for the slave owner? The stress equity is unequivocal and not in your favor.

Subconscious behavior being brought to a conscious conversation.

The sequenced equations that produce electrical energy, provide the directed energy to move the neurons of man's neural network to the members of the physical body, of what he will do next, or not do next.

This little-known fact about the human brain-inspired a billion-dollar industry with the addition of the snooze button on alarm clocks.

When we wake up in the morning, the path of least resistance is to go back to sleep. Snooze buttons are on millions of time devices, and their one purpose is to help us maximize our time until the immediate reality prevails and says we must get up. After the third snooze, the brain says, "If I don't get up and get to work, I am going to be late," and the encoded memory does not like the consequences of being late to work. The path of least resistance changed from laying there at rest to moving and doing something for a higher purpose called finances, a reward of effort.

True or false? We focus on what we expect until it is the reality of what we expected. Our reality becomes what we are focusing on in our expectations of what is next. (Because there is always a 'what's next.')

I expect to be at work on time. I focus on things that will help me achieve that expectation, like getting up out of bed, optimizing my bathroom time, and getting dressed properly. Driving the car to work would be the next focus of my expectation of getting to work on time, and the 'what's next' would be swiping the time clock. Mission complete concerning my expectation of getting to work on time.

To the point, God's brain for God's mind for God's body. The human brain is designed to focus on the expectations of receiving a reward. With this now encoded fact of wanting to lay in bed versus getting up and going to work on time, the two choices going in, knowing I could have only one answer coming out in relation to my financial obligations. Other behaviors such as shopping, excessive overeating, gambling, and smoking are in this same box as procrastinating, with the connection being short-term satisfaction. But there is no snooze button.

"I don't have to waste energy on it now because I believe I will be doing or accomplishing the task in the future."

"I will do it in a minute." (Which is hardly ever one minute, but it is easier to say I will do it in a minute, instead of exerting all of your energy into explaining why you do not want to do something right now.) Instant gratification. Deciding to have one cookie right now, or two cookies in three minutes. The path of least resistance is rooted in instant gratification. The path to that sweet tasting cookie in your mouth right now versus restraining for two minutes has everything to do with your tastebuds winning the race to your brain functions which produce motor movements in the limbs of our bodies.

So, who gets to my brain first will be the neural pathway that is turned into a bodily action, picks up the cookie, and bites. Or is it the neural pathway that keeps the members of the body from picking up the cookie?

If we look back at this simple truth from our own lives, it is experienced from the first time taste wins. Look at the little babies; taste tells them to grab all they can and give it to their mouths.

Over time, however, roughly after a few years, we start to understand the benefits of waiting for the two cookies, and that two is better than one. What really happens is that we start to understand how to reproduce dopamine release. Many and most times, our actions reflect our thoughts, including if they are compromised by outside forces or influences. It is going to cost you something.

When we hear the word 'cost,' what is our first visualized picture? Is it going to cost money, or will it cost us some time? In the human brain, there is a cost to every action produced, and sometimes the cost is that another action was not produced. If you've gathered in the wrong information, you will produce the wrong answer. Once you've gathered the right information, you will produce the right answer. More often than not, the information that we received as youth in my neighborhood was the wrong information, which produced the wrong answers in most situations. The shortcuts to the reward system could cost you time in a jail cell. The relationship to money, that neural networks hijacked on a subconscious level, is related to the power over other human beings who desire the reward as well. Sometimes, this is used to manipulate a mental decision.

"I will give you a quarter if you don't tell Momma what I did." The mental decisions with money can be seen if money can cause less pain from coming your way. Decision-making is the process of choosing alternatives based on the person's values, preferences, and beliefs. When we decide to act or not act, the brain thinks like an economist and runs a cost-benefit analysis. If the cost to act is too much, it can bias our decision-making process. It is regarded as the cognitive process resulting in the selection of a belief or a course of action among several possible alternative options.

Each decision-making process yields a conclusive choice that could, or could not, result in a final action. The human brain has a foundational principle: to be as efficient as possible. The brain does not have time to procrastinate unless you tell it that it does. To help you better visualize, the brain does not have time for bullcrap. Visualize telling yourself you've got time for bullcrap. Most people will probably say, "No, I do not." Why? Because of the importance of time, and of course, the older you get, the more this foundational principle comes into your visualized thought process. Like the sand in an hourglass, time is ticking on how to love. The brain does not have time to procrastinate while sustaining all of the body's functions. Because of this, the brain has a problem with things it thinks will be hurtful, painful, or overwhelming and tends to try and avoid things we are fearful of, or things that will require extra mental effort. The brain thinks, "I am going to assign this thought to my future self because the stress equity at the moment is not worth it."

Procrastination and the path of least resistance is an emotion-focused, managing tactic of the subconscious brain. We trick ourselves into delaying or avoiding a task altogether with the hopes of delaying or escaping the negative emotions associated with the task.

"I am not trying to face my mom, knowing I got a bad report card in school, and knowing that negative moments are about to happen." It is the choice to wait, and the delay regulates the mood you are about to be in. The consent to yourself, and with yourself, of choosing to wait releases the dopamine that reinforces and solidifies the option to defer the time to your future self. Everything "I-RIE Mon," and the mood adjustment is complete. This case of least resistance directly refers to personal effort or confrontation that most people would rather avoid. Making things harder will make them less tempting.

This is one of those things they knew the human brain would do like it was designed to do, and take the path of least resistance when things are made difficult. The diabolical part is that they can label it differently when they explain it to you, and you get a certain type of visualization and the emotional attachment from the visualized word. This subconscious human behavior being brought to a conscious conversation would be the words, labels, and phrases that are purposely attached as an adjective when describing Black people.

Why in the world would you be in a rush to pick more cotton, for nearly free? The human brain will constantly want to find a path of least resistance. Our feelings will follow our behaviors. Emotional regulation is higher on the list of priorities to the brain than other self-controlled behavior, and undermines some of the other self-control efforts. Is it logical to poke yourself in the eye with a pencil? No, because it undermines you being your best self. Stopping procrastination before it starts is a big deal but understanding procrastination before it starts is an even bigger deal. Teach me to fish.

One of these choices will keep you from acting and responding to the other choice and receiving the consequences from those manifested actions. The irony is, if we had been taught this information at an earlier age, it would have become like second nature in our subconscious thought process.

Procrastination and the Path of Least Resistance

The brain gravitates towards the most suitable pursuit in doing the least. The brain will always take the path of least resistance unless it is forced to seek another path. The brain will seek out the necessary information to do a task until it finds the shortest method that it can. It will take an authoritative or parent figure to reactivate your brain to seek more information and conclude a different course of action. Parental guidance as in action during our formative years, and taking away the shortcuts, so that the task is done with acceptable results. (Cleaning your room is not shoving everything under your bed.)

Understanding the benefits of learning about the delay of instant gratification at an early age will allow the spiritual benefits to begin to unfold, right here. God, aka Love, and the Energy Force of all Creation is the Sovereign Authority over all creation. God breathed a measure of Himself into the Body of Man. Is it such a leap to see that little sovereign authority transferred over into the makeup of Adam's brain and his thought process, which produces Adam's actions, producing Adams's habits? This, as a result, produces Adam's character and his desire for the reward. Money, Power, Respect. Is it such a leap of thought to see those same sovereign authoritative characteristics transferred into woman, as she is taken from the side of man? And now with the individuality of Eve's thought process, that produces Eve's actions, that produces Eve's habits, and produces Eve's character, in her desire for the reward as well.

We should be helpers of one another, which provides the biggest reason as to why an antagonist of this book would want the reader to believe in the words, labels, and phrases that divide and cause confusion, versus words, labels, and phrases that encourage us to unite and cause oneness.

There is only one way that a man and a woman can submit to each other, and that is to submit to the feelings of the way God loves them separately, and then respond to each other with those same feelings derived from those visualized images. Visualized images of anything and everything that points to the Love of God, The Love we feel from God. If our actions, habits, and character do not express, reveal, or show ways toward Love, we are wasting time.

Procrastination and the path of least resistance play an important role in the decision-making process, simply because it is time-related, in how we respond to the issues of life.

"Violence in everyday behavior, violence against the past that is emptied of all substance, violence against the future, for the royal regime presents itself as necessarily eternal. We see,

therefore, that the colonized people, caught in a web of a three-dimensional violence, a meeting point of multiple, diverse, repeated, cumulative violence, are soon logically confronted by the problem of ending the royal regime *by any means necessary.*"
—*Frantz Fanon, Alienation and Freedom: Part III, Ch. 22, "Why We Use Violence"* - **1960**

"I was not the one to invent lies: they were created in a society divided by class and each of us inherited lies when we were born. It is not by refusing to lie that we will abolish lies, it is by eradicating class *by any means necessary.*" —*Jean-Paul Sartre, Dirty Hands: Act 5, Scene 3* - **1963**

"We declare our right on this earth to be a man, to be a human being, to be respected as a human being, to be given the rights of a human being in this society, on this earth, on this day, which we intend to bring into existence *by any means necessary.*"
—*Malcolm X, 1965 [3] by Malcolm X at the Organization of Afro-American Unity Founding Rally on June 28, 1964*

Brooklyn Hustle

"By any means necessary" The most iconic four words of that speech; words taught as if from Malcolm, but, as you can see, the visual image came from earlier times.

I was five years old, and in the primetime of my formative years of creating neural pathways, when Malcolm X spoke at the Organization of Afro-American Unity's Founding Rally. Truth be told, with that as a foundation, I became a bad boy that did not get caught often, by doing whatever was needed to succeed, by any means necessary. More subconscious behavior being brought up for a conscious conversation: I wonder (and I'm sure many others have also) from a subconscious level, based upon Malcolm X's articulation of those words, what specific visualized images did he recall? What images did he attach to the impactful words, **"By Any Means Necessary"**? Why did these words become so popular? Because of the connectivity. The words he used were intended to connect to the impulses that Malcolm wished it would connect to. He was likely recalling from his own memories and the memories and history of America, paired with what is revealed in the tapestry and fabric of this great nation. His recollection made him visualize instances where people of 'good genes' had done anything and everything necessary to get the money.

Everything from 1619 to the ability to drop bombs out of the sky in Oklahoma resulted in the power that brought about great respect from others trying to take the shortcut as well. This is America: Money, Power, and Respect.

Any means necessary was done unto Black people, even to death, to get Black people to mentally understand that the antagonists believed they possessed the right to kill you if you did not do what they said. The same still occurs today, as this is the only rational or irrational reason that could exist when you use your common sense of deduction to keep the unfairness fair in their favor: *by any means necessary.*

Use your own visual image to fill in the blanks, and you can see the reflection of those chosen words more clearly. It is no different than the words, labels, and phrases 2Pac gave to the meaning of "T.H.U.G .Life" arising from The Hate U Give. *(Shakur, Tupac. THUG LIFE. 1994.)*

I say by any means necessary, and I mean exactly that. Then you get the power-- "Money, Power, and Respect," helping you to eat right "Money, Power, and Respect helping you to sleep better at night. Money, Power, and Respect, "What you need in life, to see the Light, it's the key to life. Money, Power, and Respect,"

(-Inspired by the Lox, featuring Lil' Kim, Money, Power, & Respect. March 17, 1998.)

Man wants power and respect so badly because of his remembrance of the sovereign authority he had when he was in union with God. I liken it to Adam's fallen thought process and replacement for the type of 'juice' he had when he communed with God.

Manifesting your thoughts into actions will cost you something, starting with the mental energy exerted and ending with what is manifested for the world to see (in your actions) what your hands do, where your feet go, and what your mouth and body expressions communicate.

Because we are made up of atoms, everything we do is governed and stimulated by electrical signals running through our bodies. That's what you get when you mix the Breath of God with the Dust of the Earth. Pow!

Energy (hydrogen oxygen carbon and nitrogen) in an Earth suit. account for 99% percent of the Atoms inside the human body.

> Subconscious behavior being brought to
> a conscious conversation, the word 'Atom'

An Indian Sage, Maharishi Kanad, labeled it back in 600 BC and called it the Param anu meaning literally "indivisible," He thought The atom was indivisible, indestructible, and hence eternal. A Greek Philosopher Democritus, around 470 BC, believed that atoms were uniform, solid, hard, unchangeable and indestructible and that atoms always existed and they moved in infinite numbers through empty space until stopped.

When the atoms combine to become molecules, they form the elements of a single-cell amoeba, which crawled out of the waters made by God.

And let the waters bring forth abundantly the moving creature that has life.

Is it such a stretch for the consciousness of man to want to understand his surroundings in the best way that he can understand his surroundings?

At the end of the day, the consciousness of man has a consciousness of his Creator.

With that said, after reading this portion of the book, there is no doubt that you are now conscious of the God-awareness inside of you. Just by the label of the word "God," you can only visualize something greater than yourself. The Conscious Breath of God when He breathed into the nostrils of Adam, His Likeness and His Image. God is Love.

-Word Check: Manifestation
To put into a physical motion of action through thoughts, feelings, and beliefs to create the appropriate equations of electrical energy, firing the neurons of man's neural network to the members of the physical body.

Now that we have this information, understanding how to stop cheating yourself out of your best life, you have to understand how to control going down the path of least resistance by knowing when and how the brain gets on this innate neural pathway. Controlling the inner snooze alarm has been exploited with the invention of the snooze button-- because the truth of the matter is you are never going to feel like it.

Think of it this way: it is why your parents made you do the things you did not want to do. Your brain wanted to take the easy way out, but your parents, with all of their wisdom of knowing what's best for their children, made you clean your room, do your homework, take out the garbage, do the dishes, and so on. If we have a serious conversation with ourselves about what we want to do versus what we need to do to accomplish what's next, we can better formulate our answers to the issues of life when we put the Dunamis Word of God that He designed to be inserted into God's brain, for God's mind, for God's body. The result is God's patience and belief in the brain that he designed and its ability to continually rewire itself until it strengthens its capacity to manifest the Word of God into actions of the body to show the Love of God, for the world to see. The connection is always there, but we must influence our actions by controlling the path of least resistance. Or we can say Grace and Mercy is the ability to constantly rewire my mind to survive my mistakes when dealing with the issues of life.

Interestingly, beliefs are typically not based on facts of the experiences but rather on one's interpretations of the facts. The path of least resistance for the brain is the path that follows the social norms and values of a society, not opposing them. The gradual change in the effort of responding causes a change in how the brain interprets visual input. Is it hard to believe that our daily decisions could be modified through a well-thought-out cognitive curriculum that includes some real talk psychological strategies, which have not been taught before? So here is the platform to articulate your neuronal thought process and what is going on in your head as the stories go back and forth, between the conversations of your conscious and subconscious mind, while you deal with issues of life.

What paths of least resistance did you have to overcome? Governments deliberately use this strategy to change how the public perceives a situation. If something seems too hard, it does not hold the same value as something easy. The path of least resistance will come into play. The end game that society teaches us is that time is money: the reward in the natural sense. But as we understand at a later stage in life, the real game is, "Time is Valuable," because it dictates when we do what we do. The brain will give you five seconds or less to make your case to do something other than what it thinks it should be doing, with efficiency in running the human body. The Spirit Man Speaks

Curriculum is written with the purpose of bringing subconscious behavior to a conscious conversation. It is an alternative to doing what we always have done regarding understanding the brain's mechanics of procrastination, and the path of least resistance, meaning we will continue to get what we have always gotten.

"The definition of insanity is doing the same thing over and over and getting the same results but expecting a different one." - Albert Einstein

It's plain to see the results that will continue to happen from this lack of information and this lack of knowledge. It is this lack of knowledge that causes my people to perish. It can lead to some positive ideas perishing, and positive intentions to do something within a certain time frame turn into not happening that way for various reasons.

Procrastination is not a synonym for laziness, but this is one of those words and labels with negativity attached to it, and then attached to Black people for them to believe. If you are told something that is not 100% true and believe it, by default, you are believing a lie, and will receive the benefits or consequences that come from that belief.

Procrastination and How to Self-Regulate Internal Mental Distractions

There have been many known predecessors that lead to other bad behaviors under the lack of self-control umbrella, such as overeating, a gambling problem, or overspending,

Dread happens, fear, being afraid of, worrying about, being anxious about, being terrified by, shying away from, trembling at the thought of, cringing from, shrinking from, having cold feet, anticipation with great apprehension, and sometimes producing uncontrollable anxiety. This can all often cause wild, unsensible behavior. Why would you do that? You lock into one thought and don't give another choice of thought to get back into something else.

This path of focus on focus goes down to our very core to focus on our conscience and on God, our Creator. No matter how much you have or don't have, you still have to have God the Creator as your hope and sustainer of life. So, once you get to the point where something challenging is happening in life, you've got to take it, and keep on moving, because nothing has really changed-- your life, and my life, are still in His hands.

The subconscious mind and the conscious mind are having it out—Panic Time. The subconscious says to the conscious mind, "Get out of the way, I knew what to do all along, but you kept telling me that I had time to do it. You tell me time is running out for real this time." The subconscious behaviors take over the manifestation of what needs to be done.

Procrastination runs the thoughts of doing things that please you at the moment.

Don't wait for an opportunity. Create it! It is the God in you! When you wake up with determination to be your most authentic self, you go to bed with the satisfaction of your most authentic self. Do something today that your future self will thank you for.

Learn how to delay with the response to the question, 'Do I want that extra piece?' grouped with the knowledge of how the impulsive brain works when it is doing its hasty checklist, and understanding how the 5-second rule of brain activity can keep the rest of the body safe.

It is more subconscious behavior being brought to a conscious conversation, and dealing with ways to reverse engineer things that tear down human confidence and anything that tries to make us go down the rabbit hole of anxiety, by controlling the story you tell yourself.

If you tell yourself you've got a bullet head, you will start noticing all the sharp edges in your head that make it look like a bullet. Take control of your mind by understanding the story you are telling it. Tell yourself a story that will boost your confidence, fight anxiety, and increase productivity as you visualize an outcome where you get to win.

Win at becoming the person you want to become! When you see yourself becoming that which you want to be, you create neural pathways to become just that, by eliminating all areas of resistance that keep you from your desire to stay focused. Pay attention, and don't get distracted. The focus on focus-on-focus method is remembering you have to focus on what you have programmed your mind to focus on, if you have told your mind that you want to.

Be mindful of the things you already know, knowing that the trick is recalling what you know correctly to focus on the research you have already done. This is the box to put stuff in that you have already figured out. Recall your bad moments, and relive your *what-the-h*ll moments*, when you concluded that you messed up, and there is no one to blame because you have come to the reality that you have free will to choose, and it was your choice to choose. If you were in a situation where you gave up your free will because you did not want something to die or suffer hardship, the path of least resistance was in full effect; especially when you gave up your free will to others.

This is the subconscious behavior that I believe needs to be further discussed from this point of view. I'm no doctor, but I've been around to see a lot and to have a very good sample of the population in my brain. I've come to a conclusion. Visualize being on crack cocaine for three days, walking out into the morning sun, and having to walk across town because you have spent your carfare on drugs. You are walking home, and you see the city is moving on without you, just waiting to see if you are going to fall off the edge; in a city that believes your failure will make it easier on everyone else: the real thought process of society.

But it was these bad moments, of recollection, that did the trick. Using God's brain, for God's mind, for God's body, can absolutely turn any bad situation into a more positive one. The flesh lusts against the spirit, and the spirit lusts against flesh. For the most part, most human brains will take a minute to think and rationalize with the subconscious before concluding and taking action.

When we procrastinate, we add the element of instant gratification to our thought process. The notion is, "Let me take a look at this right quick, and I will be satisfied." This is a subconscious move being brought to a conscious conversation. Instant gratification is very closely attached to the "shortcuts to the reward system," or the thing that knocks people in to another "creative" gear and it's called "time", and it's called "get 'er done."

When a certain alert goes off in our heads, and we realize the "rubber has met the road," we either have to crap or get off the pot. This means we have to perform the way we knew we always could when this moment came or realize we are running out of time. Mr. Panic sets in, and whatever thoughts of delaying and having a good time are quickly and very suddenly thrown out of your thought processes. Because you use your free will in correlation to the time, or the time you think you have, only to realize now you are running out of time to complete the task (i.e. study properly for a test), you end up underachieving and fail to reach your potential. This eats away at you over time, leading to elements of regret and self-loathing.

True or false? The reason you would want to avoid something is that you are afraid of the consequences.

Your foresight predicts an anticipated emotion that it does not want to feel. Fear is subconscious behavior being brought up for a conscious conversation. Why are we fearful of failing? Why are we fearful of being embarrassed? Why are we fearful of being rejected? Why are we fearful of being in pain or being hurt? (spiritmanspeaks.com)

Things get twisted when one becomes fearful, causing procrastination and the path of least resistance to step in.

> The fear of failing.
> The fear of being embarrassed.
> The fear of rejection.
> The fear of pain.

Procrastination is having the knowledge of an impending deadline for a task, yet still continually choosing to avoid starting or finishing it. This could be habitual or intentional, even with the looming possibility of negative consequences due to the delay.

Procrastination and spiritual reality

Everything from above, and avoidance in doing tasks that need to be done before we leave this Earth fall under the heading of procrastination and spiritual reality. At the end, on our deathbeds, I can imagine that the panic of hoping we got it right sets in. The panic button that is somewhere on the left side of our brains is pushed, and we do what we do to accomplish that task, while pushing the present task to the back burner.

Understand why we lose focus from being our most authentic selves: we are more likely to make mistakes in dealing with the issues of life when our conscious mind and subconscious mind are not in agreement In our most authentic state, we know how to think most efficiently without the added steps of putting on the mask that societal norms have taught us to wear.

Lessons from the School of Hard Knocks is for creating new neural pathways that are more stimulating for increase than the ones that you use right now that are based on societal norms.

It is my conclusion that a curriculum can be created using GBGMGB that will show you how to do just that.

While shutting down the neural pathways without ever knowing about the setup from the beginning (being born in sin and shaped by others who were born in sin), we are shaped in iniquity; resulting in the negative energy of unfairness, injustice, prejudice, discrimination, and bias that surround us today.

When we do not know the whole truth of why we think and how we think, we allow others to think for us. Do our young kids who are susceptible to getting left behind get the right type of tools to change their behavior, or would that lead to catching up? Our immediate surroundings nurture our neural pathways. What are they?

The new curriculum is about gathering the information of our seniors to collect the data and the stories of how they got over/overcame a lot of things that society has shown, and taught them. This is serial data of how they were able to shut down the neural pathways that lead to being bad or disruptive to the set rules of the world as a community and the things they focused on to create and grow new neural pathways by thinking outside the box. Through this collected data, we will have stories that are effective for visualization and show the consequences of procrastination and the path of least resistance.

My question is, would this have a balancing effect, knowing with a certain certainty the end from the beginning, leading to a visualization of the end from the beginning, and stopping it before it starts?

Stopping the bad, faulty information from reaching your brain center that opens the doors to negative emotions and feelings.

Stopping the path to negative manifestations of what the mouth says, what the hands do, where the feet go, and most importantly, what the mind plots from that bad information.

Knowing what is going on in your brain adds clarity to see the most positive and logical choice to create the most positive outcome. for your future self.

When you have the moment of, "I don't feel like doing (XYZ) right this second," and that you have the desire to give your mental focus a pause; remember you have already concluded value on something. You have already deemed it important enough to do to achieve your goal.

The Path of Least Resistance and the Relationship with My Significant Other

The brain desires to stay in a calm state and let LOVE do what it was designed to do. Love is patient, love is kind -- and then the submission to each other is how it's supposed to happen.

But first, better neural pathways must be created to clarify this twisted place and where it gets twisted in a marriage. Now, we are talking about some real oneness here on Earth, which leads to long-term marriage. Let's identify where the twisted gets twisted in this subconscious behavior being brought to a conscious conversation.

"Yes, honey, you are right," when subconsciously you are saying, "Whatever you say, honey, I just don't want to fight or be at odds with you anymore." If you're not saying your truth, you are acting out a lie in regards to being your most authentic self.

True or false? It is better to say this statement, even if it's not true. (If you are not sincerely in correct understanding, but you are in the path of Least resistance/avoidance)

This is a product of understanding the neural pathways that are created and thickened in the relationship between husband and wife.

Were you just saying yes because it was the path of least resistance-- perhaps to you getting back to taking a nap or watching the football game? Or, was your reasoning for saying yes your forethought, and you absolutely did not want to do anything that would decrease your chances of that oneness with your Eve, trying to get as close to God as two humans can get.

God is going to make you laugh.

Why do lovers call out to God "O God" when they are intertwined as One? #Drop the mic (spiritmanspeaks.com)

The mind thinks, "Even if I disagree with you, I will agree with you so that I can get close to the crazy amount of dopamine released." When the male brain thinks about the possibility, he will exchange other valuables for it. I mean, it is the oldest profession for a reason.

When one concedes his intelligence of something to prevent conflict, oh, what a slippery slope we have stepped onto!

A lie can be running around the house or swept under the rug, but it does not change the value of the truth. We can have peace today, and the path of least resistance will still do what it was designed to do. We receive consequences of not knowing these facts as we respond to the issues of life. The root cause, if you will, is the place where the "twisted gets twisted at," and the place where the "flim flam" happens. What regulates your heart and mind regulates the path of least resistance to what you want to do next and, most likely, what you will need to do next...because there is always a what's next.

Close your eyes and ask your own spiritual understanding; "What is his or her focus, and what does he or she expect to receive for overcoming the mind of sinful Adam?" When subconscious behavior bypasses conscious thought and inserts weighted variables into your belief system, the decision-making process is accepted by you as a positive plus thing to do.

These are in-a-split-second decisions. We are at a disadvantage when we do not understand, for whatever reason, what those variables are, or what those variables can do, positively or negatively, in our decision-making process.

The conclusion of a thought process before it is manifested into action or included in our belief system will be recalled as a belief, as we understand this neuro-science fact: it is better to know how and when memories can be recalled without your conscious mind thinking about it.

This is the space for self-reflection and agreement and confirmation that there is hope for love, in His Likeness and His Image.

Again, when we do not know the whole truth of how we think and why we think, we allow others to think for us.

The curriculum will help our young kids, who are more susceptible to getting left behind, to understand to a better degree than they understand now regarding whatever is going on in their own brains. The SMS curriculum identifies with more clarity and more often the right type of tool to use to create the right type of pathways needed to overcome the systemic mental limitations that are in place.

God's brain, for God's mind, for God's body. Visualize that. The Word of God that needs to be inserted into the thought process will change the neural pathway of the individual's brain. There are scientific facts that show the brain has the ability to rewrite and rewire neural networks. It aligns up very well with a way to renew your mind to that of the mind of Christ Jesus.

Is it such a stretch to believe, if God made this body for His Glory and Purpose, that He designed within this brain for Adam the cognitive capacity to produce the manifestation of

His Word, or a living display of His Word, so we could choose Him, which is to choose Love?

We have the awareness that we have a Creator. Still, we do not want to consciously follow that awareness. That is the trick of the antagonist of this book: to disrupt that awareness and the consciousness that produces the thought of positivity; to block the consciousness that produces a recognition that there is a God, the Creator of all Heaven and Earth, and I am a creation of the Breath of God that made Adam a living soul. #Power of God

Don't lose sight of the fact that the basis of our love is God, and His Love towards us.

In fact, it is because Love is from God. True Love has its origin in God and flows from out of God like the Breath of life.

God made the Earth by His Power, established the world by His Wisdom, and by His Understanding, stretched out from the Heavens.

God will never tempt us more than we can bear. He will with every temptation provide a way of escape. God cannot lie; then that means He had to make a design in the human brain that would figure this out, using the Free Will He gave us. #How 'bout that!

Nothing new under the sun. Free will has two choices. Think about it. All day, every day, there is the free will to choose between two things. Two thoughts go in, and one result comes out.

We learn how to stop it before it starts by visualizing the probable outcome. One thought going in would be, "How good this [whatever] will feel if I choose to partake." (for example, drugs or other bad habits).

The other thought going in will be, "How horrible I have felt in the past, and will definitely feel, once all my money is gone." (The subconscious conversation between me, myself, and "I")

The subconscious behavior of dopamine and drug related stimulants that alter the original design of how dopamine was to be released into the brain, which can alter the story you tell yourself.

By identifying the place where the twisted gets twisted at, one creates a starting point to overcome the thought processes, frequencies, and electrical impulses that will be turned into real motor functions of our bodies: again, what our hands do, what our mouths say, and where our feet go.

To the Point

Test me, ask a wise man if I'm telling the truth. What if you were to look back on your life and see how often things were good and would have been even better if you had possessed the correct information to manifest your thought processes into actions. Different thoughts would have produced different equations, which would have produced different conclusions, with the end game being the ability to choose more clearly and more often a mindful state of God-consciousness while on this Earth.

With that conscious reality, you can create new neural pathways that overcome the old pathways of fear and negativity.

My neural pathways that are attached to my consciousness are also attached to my consciousness of being a man, and the fact that I do have a Creator who is greater than I.

When we use our free will to visualize what these words mean and conclude mentally that we can do anything we want to do (except what Adam could not do), we have the ability to be reconciled mentally to the thought process of God.

It is extremely hard to progress if you don't fail. How can you win if you don't know that you lost?

It is doing the right thing at the right time, and only when you realize that you might have good intentions. As we go through life, we start to learn what to do with our time, and our minds start to correlate that there are only 24 hours in a day.

Twenty-four full hours that I get to use my free will to choose, particularly because I have a better understanding of how the brain works.

How can we know what Heaven looks like if we don't know what Hell looks like? In my opinion, Heaven looks and feels like Love. Hell must look and feel like hate.

Procrastination and Colored Peoples' Time

Ever since the beginning of time, the foundational principle has been the mental state of the human being.

Suppose the story you eventually tell yourself equates to the members of your body desiring to manifest Love. In that case, His Likeness and Image will clearly show the folly of not embracing the common denominator of us all. LOVE.

Through this collected data, we will have stories that are effective for visualization and show the consequences of procrastination and the path of least resistance.

Life's lessons from the School of Hard Knocks have taught me that knowing with a particular certainty "the end from the beginning" can help in the area of stopping it before it starts.

Knowing this information about procrastination and the path of least resistance is designed to do just that.

Knowing with a certain certainty that if you believe certain words, labels, and phrases, and their visual imagery, it can lead to subpar emotions and affect your subconscious behavior without you consciously knowing it.

Most of us have no idea how many breaths of lung contractions we have had in the last five minutes -- we have not thought about not one (until you just read this). The same goes for subconsciously doing things that your conscious awareness is unaware of, but nevertheless, are still happening.

By stopping the bad, faulty information from reaching your brain center that opens the doors to negative emotions and feelings, you are also stopping the visualizations of everything that leads to hate.

It is necessary to have a strong enough neural pathway that says, "I can do all things, because of the Love that is in me, which is the God that is in me."

Here is the real argument. In my belief system:

- certain things work perfectly fine for me, and I firmly suggest things go a certain way. Anything else would be very hypocritical for me and my brain,
- The Bible tells me I can do all things through Christ Jesus.
- I have no Heaven or Hell of my own to assign anyone to,
- I have no control over your free will,
- I have no desire whatsoever to control what God, the Creator of all things, gave humanity,
- Free will is not an illusion. It is a tool God designed for The Spirit of Man,
- The Breath /Power that God breathed through the nostrils of Adam is to use, to get back to him.
- God made this thing simple, and it is man who complicates it,
- I am going to manifest what I believe. Love, in His Likeness, and in His Image
- How to motivate in the moment is based on the story you tell yourself in the moment, creating a stronger emotion than the one you are in at that moment,

- There is a thin line between love and hate. Which side you end up on depends on the story you tell yourself,
- Knowing where that story comes from. and why, and that the visual image you will create is from your recall of memory.
- By recalling a positive outcome with a reward or recalling a negative outcome creating bad feelings like anxiety and emotional distress, you could encourage a need for emotional equity to be calculated.

Is it a good thing to understand the benefits of learning (at an early age) about the delay of instant gratification?

True or false? Procrastination is me focusing on my feelings in the short term and undermining myself in the long term.

When we feel dread or apprehensive about a task at hand, we are more likely to procrastinate to feel better. Avoid the negative task; avoid the negative mood.

With no statistical data at all, I would wager that knowing about these behaviors at a younger age would produce a better foresight in one's earlier years.

Although the choice to delay the task offers relief to our present self, it is ultimately our future self who will experience a decrease, as it is our future self that will still have to complete the task.

The problem is the reluctance on the part of our present self to do what is necessary now, with a little help from the innate program in the brain called the path of least resistance.

The path follows the laws of least resistance that are in nature, which begins with the story you tell yourself. These are the stories that you tell yourself to make yourself feel safe and secure in transferring the update to our current belief system, and that we really do believe we will "Feel more like it tomorrow."

When we rely on our present to predict our future, we can feel positive today at that moment, but we also predict our future self will feel positive as well.

Who would predict that their future self will feel negatively or fail at a task? Isn't that crazy?

Everybody, raise your hand if you want to set yourself up for failure.

Now, raise your hand if you want to avoid setting yourself up for failure.

Everybody raises their hands and understands the positivity of knowing this information versus not knowing it.

Our feelings will follow our behaviors. Emotional regulation is higher on the priority list to the brain than other self-control behavior, and undermines some of my other self-control efforts.

Knowing these subconscious behaviors will illuminate ways to prepare for decision-making, shining light on our current thought processes.

To reiterate, beliefs are typically not based on facts, but rather on the experiences and interpretations of the facts.

The path of least resistance for the brain is the path that will follow the taught societal norms and values of the society or social group that they are not in opposition to.

Take, for instance, the normalcy in wearing a face mask in public now. If you don't believe it, just go into any grocery or department store and see how many looks of, "What's wrong with you, going against what we have all have concluded that you should not," in reference to not wearing a mask. Therefore, it often dominates, leading you to engage in activities that make you feel good not to have to go against the grain that societal norms have set.

It will be significantly less multiplied by the one thing we all have in common, the Breath that God breathed into humanity.

By making things seem unattainable, the brain will say "fuhgeddaboudit." You can see the correlation between doing what you have to do to be on time versus doing what you want to do and still being on time.

There possibly is a time and place to move at your own speed, but to this point, CP Time cannot continue to be swept under the rug after this.

The SMS curriculum is an alternative to doing what we have always done with regards to understanding the brain mechanics of procrastination and the path of least resistance; which means,

"If we continue to do what we have always done, we will continue to get what we have always gotten." This is a prime example of the human mind believing what it was trained to believe, then perpetrated so much that it became a normality to be late.

Chapter Nine: Ninth Grade

Love, Sex, and Oneness

(A conscious conversation about a subconscious behavior would begin something like this: what's up with Adam and Eve, and why, over the history of time, men have gotten in so much trouble for "sticking their PePe's in the wrong place." Then I would ask the question of why it's not talked about from the pulpit, when it's such a big problem that's leads to a "whole "lot of negativities.)

And what you did not learn in church

THE SPIRIT OF Man is willing, but the flesh of man is more willing. The hardest knocks upside my head come from an incomplete understanding of love, sex, and oneness with the Creator of all Heaven and Earth.

As I think about it, love is the feeling that our brains crave the most, because of that dopamine drop. It wants to be loved, and it wants to give love. It wants to *be* love, and that's the sweetest dopamine drop there is.

Everybody wants to love and be loved, and everyone wants to be connected and have a sense of belonging. We want to love, have sex, and be one.

We want to be connected in some way to another human being.

We desire these connections because of the original neural connection to another human being through the umbilical cord.

As you think about it, is love the feeling that your brain craves the most?

Everyone needs somebody to love, or some other living soul to love, even animals.

It is in these cravings of love that the human brain will produce equations with varying degrees of feelings and logic interjected into the spectrum of final thoughts to be manifested into your actions for the world to see.

It could make very little sense, but feel right. Or, on the flip side, it makes sense, but I feel afraid to do it; the conscious awareness of subconscious behavior, logic, and feelings of what we decide and how we decide.

Love, sex, and oneness have put most of the knots upside my head, but I am grateful that as I understand my innate power more and more each day, the knots are a lot less frequent and not as big.

We often search for love in all the wrong places because life starts us out with us looking in the wrong direction.

We start out looking outward for love, as opposed to looking inward for love. By default, this limits the power within each individual because no new neural pathways are created within to be manifested outwardly.

The word, label, and phrase we call love is the same label and phrase we have in our belief system as the energy source of all living things. #God is Love.

With that said, because the word *love* is named by the mind of man, it can also be manipulated by the mind of man.

More subconscious behavior being brought to a conscious conversation.

How many people have been hurt in the name of love where it was used in a mentally manipulative way, or used as a derivative of the original feelings we came into the world with? How many people love a Snickers® bar the same way they love their mother?

Logically, we can agree that the love of the candy bar is a derivative of the feelings we have when we are talking about our mothers. We will do a lot more to protect our mothers than we will to protect a candy bar.

How many young men and boys have stood up to their dad, stepdad, or father figure? How many got punched like a man in response, or got the mess knocked out of them by him?

What about interfering in a domestic altercation with your mother, with you considering him a coward of a man, for messing with a woman?

Thinking of how much I love my mother brings a tear to my eye, because I recalled the episode and the emotions that came with it.

To be clear, this was the last incident that ever took place when I was around.

The umbilical cord of life between mother and child is filled with Love Energy.

The transfer back and forth for nine months of the consciousness with God is handed down to everyone born of a woman, and it starts at the original umbilical cord.

It is the neural pathway figuratively connected to God's mind and that's the same neural pathway connected to Love.

The brain is continuously trying to figure out how to love, and ways to love.

What happens in the human brain's desire to manifest that love into an action for the world to see?

It is likely to come across a neural pathway that says, "Hold up, wait a minute, don't do that." History has shown that this has an extremely high probability of taking a loss #taking an "L", and a negative return of the love I desire to send out.

Because of this belief, I will not put the power of love into the equations of my thought process that I will manifest in the members of my body. #Self-preservation.

Society's "elephant dung" has made it extremely hard for the human mind to focus on being our most authentic selves and operating in His Likeness and Image.

Subconscious behavior reveals that the flesh will lie to the spirit, and the flesh will lust against the spirit, as the spirit will lust against flesh.

Where does the flesh get its information from to lie to the spirit?

Where does the conscious mind get its information from to program the subconscious mind?

Anything outside your physical body can become incoming information to your inner body from the outside world.

The conscious mind will believe what it was taught, and programmed to believe, during the formative years and tell the same story.

This practice eventually programs the subconscious mind with those words, labels, and phrases attached to the visual imagery of what those words were taught as meaning.

The subconscious mind is designed to trust what the conscious mind puts in it and send the correct electrical signals to the body's motor functions.

The Sex Part

Think about all the songs and lyrics that have been written about love. There have been so many! "Love will make you do right, and love will make you do wrong." *(Green, Al. Love and Happiness. 1972)* Or my personal favorite, "Love should have brought your "donkey-parts" home last night." (*Braxton, Toni. Love Shoulda Brought You Home. 1992*)

Seriously, what we are talking about here is something that is not talked about enough. We're talking about love, sex, the difference between the two, the choice of love, and the feelings of sex.

Ask yourself, has anyone ever sat down with you and discussed the difference between love and sex, or making love?

The mental difference between love, making love, having sex, and masturbation to the human brain varies tremendously. (We're in the deep end of the pool now!)

The visual imagery of how the first thought of sin could arise in innocent beings is left up to the human mind to resolve.

My deduction on using God's brain for seeing what is not seen, tells me that God made this thing simple, and it is man who has complicated it.

The author of confusion, or the "antagonist in your book of life," would use words, labels, and phrases as a means to do the confusing.

What would be the best place to plant the first seed of confusion?

Correct. In the Garden of Eden.

Considering the words and labels exchanged between Adam, Eve, and the serpent, Eve could have said, that Adam said God said not to eat the fruit and not to touch it.

Eve was not on the scene when God instructed Adam about the tree in the middle of the garden.

The real question becomes, "What words, labels, and phrases did Adam articulate to Eve?" We are looking here because confusion is here.(spiritmanspeaks.com)

The craftiness and trickery of the antagonist's plan is always to find ways to use other people to present his plan into your life, by using words, labels, and phrases that invite the human mind to visualize your present, and right now, situation, as uncertain or unreasonable as it may be.

Then, for you to visualize His plan, as a better plan, a plan to increase and multiply for your future self because of the mental stimulation of words, labels, and phrases.

For the most part, the human thought process does not predict bad news (a decrease) for its future self.

Because of the mental stimulation of words, labels, and phrases, the antagonist in the book of life suits the temptation through logic and promises of increase from eating the 'apple.' #Drop the mic spiritmanspeaks.com

It teaches and presents to the human thought process the promise of increase, with regards to "If you do this, you will get that."

The focus on the *that* of the promise becomes the place where the twisted gets twisted at.

The human thought process will subconsciously visualize what those words would look like, manifested for our future self, at the very idea of increasing and multiplying from its current position.

With the stories that were just heard, a chemical reaction happens in the brain, and dopamine drops as one of the factors that would cause the human thought process to conclude that, "Yes, this is a good idea, let's make this happen," based on the good feeling you just felt, for your future self.

The aim is to make us feel discontented with our present conscious state, as if it were not so good as it might be and should be.

The unlearned mind is naturally inclined to believe the truth of every assertion until it has learned by experience that fire really does burn.

Oneness

What if it was One, like it should be, the joy of increase, and the joy of multiplication? The joy of adding value to each other's lives, to the point of helpmates-- helping one another, to be pure love. That's the journey that all relationships are on.

Selfishness must be absent for love to be pure, and the first step is to love God with pure Love.

When we understand how to love our helpmates with pure love, we will find that we are drawn to love, for love's sake, and for love's feelings.

Only Love is Worthy of Love.

When a man and a woman learn to love each other in this way, selflessly, and focus only on God's excellence and beauty, the two that become one are filled with a joy so fulfilling that it becomes logical that we should continue to add value to each other's lives, and that becomes the reward.

We need no other outside influence as long as we keep the pure Love of God in our man and woman relationships. The joy that fills us as a result of loving God with pure love becomes so strong that it overflows into a thoughtful and sincere love for your helpmate and the rest of humanity.

When we add our human feelings, it will influence and have the power over the logic in our thinking and will either increase and multiply love, or the logic will divide into confusion.

Depending on your formative years' teachings, what you kept and shut down could be like still believing in Santa Claus.

Pastor the man who has raised his consciousness to all incoming equations of this type, comes down to math, a positive or a negative towards God, or away from God.

The plus shows love towards God, and the neural pathway that God created in His creation in the first place for the correct manifestation of the body. This would reflect the energy source he put in man, to begin with.

Who Are You?

The next time you see him, ask the pastor, "Who are you?" He will state who he claims to be.

He will state his consciousness of who he believes himself to be A man (or woman) of God, for the world to see, hopefully.

We are talking about their hearts, understandings, clarifications, wisdom, and interpretations of how they feel about God's love towards them.

They choose to dialogue and articulate with words, labels, and phrases, in hopes that the receiver of those words, labels, and phrases will choose with their free will to use God's brain, for God's mind, for God's body, and manifest His Likeness and Image of love, in the body that God made to give Him praise and glory.

You know, God made this thing simple regarding how to show love towards God for all to see. Love in this time period of humanity, and this part of the vineyard.

The man will always be inside the pastor, like the Good Shepherd. The love of a "good shepherd" is revealed in their desire to share how they feel, and God's love for them to all that have an ear to listen.

It is through the discipline and the immense growth of the neural pathways in the pastor's mind, and neural network that he has chosen with his free will to do so, continuously and repeatedly.

Manifesting in the members of their bodies, what their hands do, where their feet go, and what their mouth says; most importantly, what is allowed in through visual and audible input from the outside world.

From a neuroscience point of view, the filters that God designed, to begin with, begin to do what they were designed to do, allowing electrical stimulants of information that point towards love and God.

It is electrical stimulants of information that will prevent a decrease in bodily actions or manifested actions, which don't point to the Love of God.

I AM THAT I AM

The instant you read these words, the person you are right now is based on what stimulates your mind for an increase or what stimulates your mind to prevent a decrease. If you were to ask yourself the question, "Who *am* I?" I guarantee that your answer to yourself is exactly what you will be. The story you tell yourself, that you are.

Lessons from the School of Hard Knocks has knocked out some behaviors: subconscious behaviors that were not good for your body, but also created more opportunities that were good for your body.

The human mind is stimulated by two things, opportunities for an increase and opportunities to prevent a decrease.

The brain that God designed will always be moving towards love, the original Love (as in "God is Love) that it had when it could have communion with God in that kind of Loving way.

That's the kind of Love that has a lot of Power connected to it.

It is the kind of Love that gets taken advantage of.

Love in the wrong hands can damage the human psyche, soul, and mind.

Love is attracted to Love.

I can use my free will to find out if what society taught me was good for me or not. It is like believing in Santa Claus. It was a useful lie to get me to believe in something I could not see.

God can turn it around for good. The lie was bad, but because it did not kill me, the lesson was good because I am mentally stronger and can cognitively remember the negative emotions related to that issue.

I can shut it down from going down that negative rabbit hole.

More importantly, be so grateful and happy for the new pathway that is created that you don't have to continue taking losses with negative feelings stressing you out.

You can keep it moving now.

God will make a way out of no way (spiritmanspeaks.com)

The way out was always there, you just didn't have consciousness of it.

But when you become conscious of that way and have the mental focus to manifest, the consciousness of this new way of processing thoughts exists, that you did not have before.

If you had it before, you wouldn't be in the negative right now. It is the negative feeling of being in trouble, with worry, despair, and angst, vexed at a mental decision that you made early on.

When you insert this equation into your GBGMGB, it becomes the mathematical and logical equation for God to make a way out of no way.

The Curriculum

The curriculum for the brain is to rightly divide the choices that you must choose from so that the outcome equals God.

The journey is to rewire the brain to visualize God's love, to know what is good for you, even though you will not see what is good for you until you see the outcome.

This book is based on what I have seen with my own eyes, and I am changing the filter and lenses that social norms taught me to look through.

Through my lens, if you are not loving, then you are wasting time.

My spiritual lens says, without a doubt, if there is a lot of confusion in something, the author of confusion has caused it with words, labels, and phrases.

Confusion wastes time by taking your mental focus away from Love.

But this is the journey, my friends. I am just trying to give insight into the mathematical variables derived from the transition of words, labels, and phrases, from one human being to another through mental manipulation.

When words of uncertainty ("I guess", "I am not sure", "I think so") are involved in the thinking process, this grammatical equation should be applied: the *I* that is *I* is the *I* that is *We*, and the *I* that is *We*, is the *I*, that is *He*. #One Love

In my humble opinion and conclusion, the correct equation for marriage will have all the benefits of God's original plans when He created them, both male and female.

This also includes the wisdom of God, who created both for a purpose that has never changed, multiplying and replenishing, and subduing the Earth.

I did not learn this from church, so I hope I am not adding anything to, or taking anything away from the Word of God.

By looking at life from God's point of view and using God's brain for God's mind, for God's body, it is the Breath/Power that was breathed into the first man, the measured portion of God, known as the Spirit of Man, manifesting the visual intake of what the Spirit of God in man is aware of and the things that the natural man is unaware of.

I call myself, making him aware of it. I am choosing with my free will to acknowledge the measured power of God that is within, with more clarity, and more often.

This is based on seeing what is not seen: by not visualizing what societal norms taught me to visualize, but by visualizing ways to flip it positively.

The elephant dung that was taught to me in my formative years grew into the bad that I would do, that I did not know how not to do.

Words, labels, and phrases that were introduced into my conscious were the words, labels, and phrases that created the pathways that grew the way they were designed to grow and had the diabolical effects they were designed to have.

At some point, it is your choice to choose free will. It is about free will to choose and to manifest into the actions of your body. Then, everything else that is associated with it is about choice as well.

Love is about choice, it is not a feeling, or an emotion.

"Choose you this day", with your free will, to represent His Likeness and His Image for the world to see.

Just like the man called Christ Jesus, the man portion in Jesus, the portion that cried out, "Take this cup from me." #That one.

The man portion that the world was able to see and interact with. the man portions that was using GBGMGB, also known as Christ Jesus.

Is it better to talk about the man portion of God, as an example? Or is it better to talk about the God portion of Jesus as an example, the Breath of God, and the manifestation of Christ Jesus?

It is the full Power that is God, in human form.

The world identifies with that image form as the man Christ Jesus.

It gets me to love, and I love you, so I must tell it like I got it in my grammatical equations that produce actions that show the God in me.

In my mind, the next thought that comes up is Christ Jesus as the best way to represent God's Breath, the Energy Source; something you can feel but cannot see unless it is seen in another human being.

When I use my free will to tell myself a story, I tell myself the first man is an earthy, bad little boy who grew up to be a bad "donkey part" man, whose formative years took place in Brooklyn, New York

I tell myself that the second man is experienced. Through trial and tribulation, I am still here in my right mind.

When I say the name "Jesus," I know I can get some strength to just keep it moving; to keep it moving in a positive direction with a desire to increase and multiply love, as I deal with the issues of life.

God's creation of Love looks exactly like it feels.

Love is purely an action that you choose to manifest in your flesh.

It is using mathematical equations to filter emotions before they are manifested in your actions, which will be an image for the world to see.

The image seen either will, or will not, connect, to the image of the Love of God in the beholder.

Chapter Ten: Tenth Grade

Zeros and Ones

The Creator made this thing simple, it's Man who complicates it.
0x0 =0, 0x1=0, 1x1 = 1

THEN GOD BLESSED them, and God said to them, "Be fruitful and multiply; fill the earth and subdue it; have dominion over the fish of the sea, over the birds of the air, and over every living thing that moves on the earth."

God blessed man, and God blessed the woman with the same measure of POWER. He bestowed upon them both a mental understanding of how to transfer that energy to be fruitful and produce more Love in His likeness, His image.

He bestowed upon them the ability to multiply and fill the Earth and replenish and increase by creating the babies born in His likeness. It is the measure of POWER that comes from the breath He breathed into our nostrils, after forming us from the dust of the Earth.

In my mind, the Bible was inspired by the Spiritual nature, so I believe that it takes a great deal of man's interpretation to start relying on the spiritual nature and power inside.

The Sovereign Authority that man had with God is no more, and it is possible that it has been cut off since Adam made the mental decision to "bite the apple."

The mental decisions of free will have consequences-- some good, some bad, and some ugly.

Subconscious behavior being brought up for a conscious conversation.

Lessons From The School of Hard Knocks contains my "aha" moments in life. It was not written to cause confusion, but to point out where the confusion lies.

Did you know that there are over 450 versions of the English Bible? That is 450+ versions of words, labels, and phrases that make the human mind visualize interpretations and meanings designed to make the human mind visualize a positive increase for one's future self.

The gift of Free Will to choose was designed in the mind of mankind by the Creator of All Heaven and Earth.

The Neuroscience Transparency

The facts are, when you ask yourself a question, agree with it, it makes a better imprint in the grey matter of your brain cells than if you were just listening or reading continuously. This is the purpose of my frequent insertion of 'True or False' questioning. It is for the purpose of creating transparency to what is happening in the human brain.

Science and neuroscience are words that man has created to label the thoughts generated in his mind according to what he needs them to be: for addition and subtraction purposes, in man's personal equations of manifesting an increase, and manifesting ways to prevent a decrease.

The way that God designed the mind has not changed; however, what man/woman decides to put into it, has.

What also continues to change are the words, labels, and phrases that man uses to get a visual image.

Yet and still, how the mind works has not changed one bit since God's original design (with Christ Jesus being the backdoor key to get back in, like a computer reset key).

Think of "getting locked out" as a metaphor for how to get back to your right mind after the mental brokenness, which stems from Adam's critical mental decision, and from which we have all suffered.

Even with the back door key (Christ Jesus, Free and available), you still have to deal with your own free will.

God's original intent for His design of the human brain was a desire for the mind to be free and motivated by electrically charged stimulants taken in when we hear certain words, labels, and phrases.

The original intent for God's brain was to be used to manifest God, in the body He made for His Glory and Purpose.

We can learn more, with regards to our thought process and how to use it as it was designed to be used.

With cognitive clarity, LFTSOHKS reveals the origin of where the stimulants come from, along with the effects of the stimulants and the stories they develop, full of the words, labels, and phrases that will become the story you tell yourself.

If you don't believe me, close your eyes and ask your own self about the validity of the words, labels, and phrases that I have used here.

A Conversation about Zeros and Ones

What if we could increase the percentage of the times when we have a good idea versus a bad idea?

What if we could increase the percentage of the times that we made a better choice?

What if we could increase the number of times that we have joy, versus the times that we have sadness?

What if we could increase the times that we have strength, and decrease the times that we have weakness?

What if we could increase the times that we have peace, versus the times we experience drama?

What if we could increase the positivity in all of our actions, versus the amount of time we spend producing negativity for ourselves and in other people's lives?

God made this thing simple: how to love Him, and how to love one another.

God is the greatest mathematician of all time. He is the G.O.A.T.

It is my conclusion that the challenge of the human experience is making choices that He has already chosen for us, and to choose Him.

"For as he thinks in his heart, so is he."

The challenge is continuing to remember the times we elected to make the right choice, and the right thing happened as a result of that right choice.

As we encode memories of doing the right thing, the pathways to the storage portion of your memory system will become more traveled and with increased speed.

As we increase the pathway speed and our retrieval/recall from the storage portion of our memory system of 'the right thing,' the speed simultaneously decreases the synapse connections to those old neural circuits of "the wrong thing": literally bypassing and shutting down the society-taught "elephant dung" in existing neural pathways.

As we read in Chapter 5 on Santa Claus, we can now understand, without having a Ph.D., that it is possible to change your belief system. We discussed the ability to change our belief systems, as we changed who we thought ourselves to be, from that 5-year-old child that believed in the mental magic of Santa Claus, to the 8-year-old young man that figured out they don't have chimneys on the roof of the projects.

This was one of the visual effects of my reality that I used in my decision-making process to shut down the electrical impulses that told me Santa Claus was real. The electrical stimulant that occurred when I heard words, labels, and phrases with Santa Claus in them was not permitted to pass through my Thalamus Reticular Activating System Region. The neural pathway that said Santa was real had dried up and no longer stimulated my long-term memory in the hippocampus in the form of an increase.

It made more sense to make your mom or parents the stimulated, number one source of opportunities for an increase of gifts and presents. Mom became more, and Santa Claus became less.

Neuroplasticity is a scientific term used to describe the brain's ability to evolve through reorganization and growth in response to life experiences. It is the brain's capacity to create new neural pathways.

We could also define neuroplasticity as God's design of the brain, for the purposes of putting off the old man and putting on the new.

Nothing new under the sun.

Scientists did not design the brain. They just put a name to it.

Your reason for being here, for reading this book, is to understand how to build brand new neural pathways with a spiritual understanding.

It is to form a knowledge of what's going on with the recall and retrieval system, what stimulants prompt a recall from our spiritual man's understanding, and what stimulates a recall from our natural man.

By understanding the subconscious behaviors of the human mind, the reader will be able to better identify them consciously, resulting in clarity of thought.

We can identify the stories we tell ourselves subconsciously, that will stimulate a response from the natural flesh of man or stimulate a response from the spirit of man.

We can recall and retrieve from our renewed minds rather than from our third-party taught minds by using simple, logical equations to separate the meat from the bone and rightly divide the Word of God by narrowing down the process using single-output logic gates: a 'One' gate, or a 'Zero' gate – to do, or not to do, Go, or no go.

True or false: The computer is derived from God's design of the human mind and brain, which computes and sends billions of bits of information back and forth through the central nervous system in the body.

In computer systems, the information is carried by wires and electrical parts made by man to simulate the neuropathways in the Body of Adam.

You are able to see first-hand the design, designed by the Ultimate Designer. Just as white blood cells are magnificently designed to do specific work for the health of the body, so is the TRAS - R designed for the health of the mind, by keeping the "garbage" of negative inputs out of our brains for clearer thought with more consistency of what is good for the body, and what is not.

It is through the brain, and the free will of our mental thought processes, that God gave us a way to access His measured Power within.

The journey is about using our free will to manifest positivity through the Love of God, which empowers the brain through electrical signals and stimuli.

The electrical stimuli are manifested into mechanical actions by the members of the body...

When considering zeros and ones, for God's brain, God's mind, and God's body, assume God represents the One, while zero represents anything less than God.

Zeros and Ones

The brain will resolve our logical choices by narrowing down the process to a single output from a logic gate, of zero or one.

Put simply, computers are based on logic gates, using the binary code of "0's and '1's".

The logic gate must produce a single output that is wholly determined by the sequence of inputs.

When we program our brain to have the Dunamis Word of God as an input into a logic gate, the spiritual theory of GBGMGB will be easily allowed through our TRAS-R.

An accurate application of Zeros and Ones in our decision-making process will always get the Spirit of Man back on the right path and back to God. (spiritmanspeaks.com)

There is data in your memory of the times you did not choose to manifest the Love of God, in your actions and remembrance of what happened and the times you did, choose to manifest your God-given common sense in your decision making. #using the good sense, the Lord gave you.

There is built-up wisdom in your memories of things you have already figured out: an assortment of things, like not touching a hot stove knowing it will burn you. Once we have learned this, we move subconsciously around a hot stove.

The brain adds things like this up automatically, and the thought never even makes it to our conscious mind for calculations. It creates a rightful divide between the consciousness of the Natural Flesh of Man and the Spirit of Man that is dormant until called upon out of subconscious thought to conscious thought.

Understand that it is always there. It is like, understanding that God has already provided a way of escape from Adam's original thought process.

With our free will, we are able to program the TRAS-R to do what they were designed to do.

We can program them to build new filter systems that point to the Love of God.(And All the people said Amen …It is as simple as that.)

The "rEVOLution" is about what happens if we consciously program the billions of neurons and millions of logic gates into a sequence of brain processes that produce the quickening, the speed link of knowing something without consciously thinking about it. Imagine now with the Dunamis Word of God as the accelerator for the quickening.

-Word Check: Dunamis
Greek philosophy meaning power, potential or ability…a root word for " "Dynamite", describing the Power of God that is in His Word.

This gives Power to grammatical equations of words, labels, and phrases to be used in the Brain, Mind, and Body in manifesting His Love for the World to see.

The relationship between inputs and the final output can all be graphed to show the tendencies from which we get our conclusions. Our inputs from choices and memories signal to us that we are going to take a loss and suffer from low vibrational emotions, or a less-than-one feeling.

Natural Man or Spirit Man

The subconscious logic now becomes the neural "tract" speed of delivery from the natural flesh of man or from the Spirit of Man

The neural stimulant that travels with the speed of Usain Bolt, and gets there first to be recalled and retrieved from our memory's storage rack creating 'the story' we tell ourselves, and manifesting in our bodies of what we are about to do, and how we are about to do it.

The zeros and ones of the human experience are the basic building blocks of what is good and what is bad.

The zero and the one represents the two inputs of any logical equation of our choices: choices A and B. The zero and the one come from the stories we tell ourselves, based on the knowledge, wisdom, and understanding of the choices we have made in the past.

The flesh of the natural man is one variable, and the spirit of man is the other variable.

In this discussion, we will label the flesh variable "0" and the spirit variable "1."

Zeros and ones are the mathematical way to apply neuroscience to the decisions we make.

Zeroes come from third-party programming, and ones come from our innate spirituality that God is Love and the voice inside that conveys what is good for the body and what is bad for the body.

When we have mental stimulation, such as going to church or reading our Word, we have an opportunity to encode that Word into our long-term memories.

This is good for our physical bodies, as well as the Body of Christ and the church.

Choosing a positive path leads to more positivity (ones) and creates positive neural networks with a positive train of thought.

Subconscious behavior being brought to a conscious conversation

Subconscious behavior, known as "Confirmation Bias," has relationship to the story that we tell ourselves, that manifests into the members of our bodies for the world to see.

From a 'Brooklyn Hustle' Point of View

One dollar is better than zero dollars.

With that same mathematical equation, we can see that one dollar is better than 75 cents, and that one dollar is better than

20 cents. The same equation will tell you that one dollar is way better than one cent.

[Figure 2 -3 Less Than $1]

Penny	Nickle	Dime	Quarter
1 cent	5 cents	10 cents	25 cents
1/100 of a dollar	5/100 = 1/20 of a dollar	10/100 = 1/10 of a dollar	25/100 = 1/4 of a dollar

Dollar
100 cents

Put plainly, one is better than less than one. The Oneness of God is a better choice than choosing anything less than the Oneness of God.

Mankind has the free will to choose a zero, or one, a positive over a negative, and love over hate.

As we read, the Word of God becomes encoded into God's brain that He designed for His Glory and Purpose. We can now begin to see how the impossible of man's thinking can be possible with God's thinking.

A man has the free will to choose the human experience that he desires, and free will is about choosing between this or that.

Zeros and Ones for harnessing the power of choice.

Zeros and Ones for accepting how to navigate beyond the negative.

Zeros and Ones for harnessing the power of redirecting oneself.

Zeros and Ones for the best forethought that you can possibly forethink.

Zeros and Ones for choosing the positive and seeing the light, everywhere, in all things and all experiences.

As your mind adds and subtracts from the original intent, you are now actively manipulating certain information to get to where you need to go to keep progressing towards your desired goals. This is the original intent in God's original design for the human brain.

[Figure 4 - Zeros and Ones]

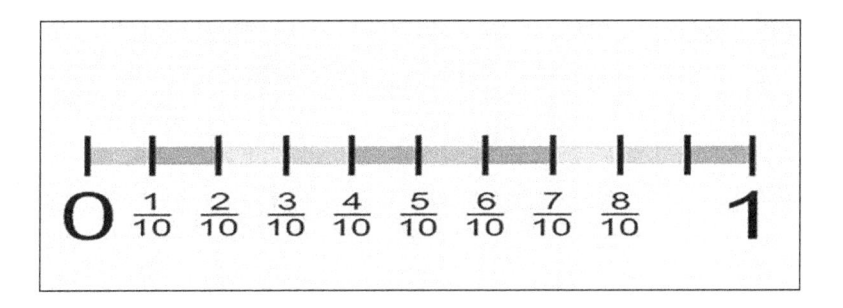

$0 \quad \frac{1}{10} \quad \frac{2}{10} \quad \frac{3}{10} \quad \frac{4}{10} \quad \frac{5}{10} \quad \frac{6}{10} \quad \frac{7}{10} \quad \frac{8}{10} \quad 1$

Use the number line and consider the idea of making yourself feel more than or making another person feel less than. When you choose to call a person a derogatory name, you are choosing a zero, and that choice will not produce a more positive direction to go in.

Correct me if I am wrong, but from my spiritual-being point of view, we are born into sin and shaped in iniquity, both of which are not good.

We come into these breathing bodies in disorder. Then we are taught and shown more disorder (wickedness, evil, vices, injustice, crime) by the systemic systems built in societal norms. #Patriarchy - is that the problem or is what we've done with it the problem?

There are many of these so-called "Kens and Karen's" out there who believe in the systemic stereotypes built into the fabric of the American thought process and programmed into them in their formative years.

The neural pathways that were created and reinforced from societal visuals that you see and continue to believe, seemingly give them an advantage in how to increase their money and power.

This also consists of using certain words that they know will affect another person from an emotional point of view.

Revealing the antagonists' unfair motives in the book of life will help stop people from inputting societal terms of hate and negativity without knowing the subconscious behaviors behind them.

Using love, we can reverse-engineer the mindset of a Karen, or a Ken, and understand where the twisted gets twisted.

Is it really her fault that she was programmed this way?

We are shaped by other human beings who were born into disorder the same way we were.

We start out in life from a two-point disadvantage, at a negative two on the number line. There will come a time when it makes better logical sense to add the Word of God to our thought processes to prevent taking a loss, and making the same mistakes, while calculating what is next for our future self.

With more cognitive clarity and consistency in our foresight for our future self, we can remember what happened when we added God's words to our thought process, or when we did not, and how it affected our positive flow in that oneness Love of God.

What if we could increase the percentage of the times when we have a good idea versus a bad idea?

What if we could increase the percentage of the times that we made a better choice?

What if we could increase the number of times that we have joy, versus the times that we have sadness?

What if we could increase the times that we have strength, and decrease the times that we have weakness?

What if we could increase the times that we have peace, versus the times we experience drama?

What if we could increase the positivity in all of our actions, versus the amount of time we spend producing negativity for ourselves and in other people's lives?

That is the challenge of the human experience --- making choices that manifest God's love for us in all of our actions towards each other.

We can label it the "human experience," but it is my hope that the many members, but one spirit, brought you to this book to help the revelation of the evolution of Spirit of Man

The best preachers, teachers, and motivational speakers all inform us that "The power is within you!" What power are they talking about? The Power of the Spirit of Man.

Take the meat and leave the bone.

We know that the flesh lusts against the spirit, and the spirit lusts against the flesh. An enmity exists between the two.

To conclude: our choices in life are based on two variables. The flesh and third-party programming being one variable; and the Spirit of Man that point to Love as the other; the disobedience in Adam's thought process vs. the obedience of Christ Jesus's thought process.

I personally believe that the end result is Christ Jesus showing up in man's Earth suit, showing us how to manifest that power in a human body, multiplying and increasing the Power of God that we were born with.

Guess how many people still believe that we only use eight percent of our brains? This is possibly because some other human being told you, or you read it in a book, or saw it on TV. It is an incorrect variable designed to keep your creativity at an eight percent level.

As mankind has free will to choose, he also has the free will to harness the power of choice and the power of redirecting himself.

We have the free will to accept and mentally navigate beyond the negative scenarios of our childhoods.

Through the simplicity of zeroes and ones, you can find the solution of what is good for the body, and what is not good for the body.

Only because man himself knows love; can he have some inkling of the power that is in Love. This is true of all the other revealed attributes of the Almighty.

Do you know why? The gap is so wide between our small, limited minds and the unlimited Mind of the Creator of all Heaven and Earth.

Whatever positive language or articulation we can use would still be negative regarding how great God is, how God thinks, operates, and exists.

This reality can be attested to when we think of how good He is in our lives, and our very next thought is, "He is better than that!"

God operating in our lives increases the positive aspects and consequences of our decision-making exponentially. That demonstrates that whatever issues in life we are dealing with will turn out for our good.

It is true that we have encoded memories of the wretched state of our mind, but we have also encoded memories of how God brought us out of darkness and into His marvelous Light.

These thoughts come from our memories of "down through the years" and remembering just how good God has been to me.

Any positive image or language must be understood as making sense only in relative terms, i.e., relative to ourselves. We are powerless, by ourselves, to produce more than ourselves.

Man's fallen state puts him out there by himself, but God's Grace and Mercy are designed to bring him back, using God's designed brain, made for the body of Adam.

The subconscious mind is designed to be influenced by positive images, like the mental image of the greatness of the Almighty as against our own human frailty.

Think of His infinite Power and Wisdom compared with our puny strength and feeble intellect. It is our intellectual awareness that God is pure spirit and omnipresent, for there is no other truth than that which is relative to the mind of man.

True or False? It seems that we fall short when we do not recall the memory of the Power that's in the Love of God that can overcome the issues of life when applied.

As man has free will to choose, the human experience is about those choices. Choose you this day, this moment of this day. Whatever thought we could imagine of how great God is would be a vast understatement to His awesomeness.

Lessons from the School of Hard Knocks has taught me that I am powerless by myself to produce more than myself.

All things turn out for the good for those who know how to use God's brand designed for God's mind and manifested in God's Body in carrying out His designed business and purpose for humanity. What is the final thought that becomes the thought that will become the story we tell. The subconscious decisions behind the choices we make, (and why we take the naps) so that we can get the rewards, are the thoughts determined by the actions of specific neural circuits in our brain.

What if you had a curriculum that taught you about the specific neural circuits in your brain, uncovering those circuits, for better clarity and more consistency in seeing the end from the beginning?

This would result in much higher mathematical equations producing the Love of God.

Zeros and Ones increase the percentage of the times we have a good idea versus a bad idea.

Zeros and Ones increase, the percentage of the times we make a better choice.

Zeros and Ones increase the times when we have joy versus the times, we have sadness.

Zeros and Ones increase the number of times we have strength and not weakness.

Zeros and Ones increase the positivity in all of our actions, versus producing negativity in other people's lives.

Zeros and Ones increase the percentage of the time we spend experiencing positive thoughts versus negative thoughts.

Zeros and Ones increase the percentage of times we feel empowered versus the times we feel defeated.

Zeros and Ones increase the percentage of times we have our hope versus the times we experience despair.

Zeros and Ones increase the percentage of time we have peace in our space versus the times in which we have drama in our space.

Zeros and Ones increase the positivity in all of our actions affecting people around us

GOD is the G.O.A.T. in Mathematics

A B C D E F G H I J K L M N O P Q R S T U V W X Y Z
is represented as:

1 2 3 4 5 6 7 8 9 10 11 12 13 14 15 16 17 18 19 20 21 22 23 24 25 26.

If true, then:

H-A-R-D-W-O-R- K

8+1+18+4+23+15+18+11 = 98%

and

K-N-O-W-L-E-D-G-E

11+14+15+23+12+5+4+7+5 = 96%

But,

A-T-T-I-T-U-D-E

1+20+20+9+20+21+4+5 = 100%

AND look how far the love of God will take you!

L-O-V-E-O-F-G-O-D

12+15+22+5+15+6+7+15+4 = 101%

(Soroye, Pastor Akinola. *Attitude & the Love of God–2012*)

Therefore, one can conclude with mathematical certainty that, while hard work and knowledge will get you close, and attitude will get you there, it's the Love of God that will put you over the top!

When it comes to my future self, I would rather be a penny over than a penny short.

The man who lives with godly expectations can never be depressed or disappointed for too long, because of the new neural pathways created and in place to look for spiritual understanding for help to input into the decision-making process.

This allows discernment (subconscious behaviors of your free will) to kick in a lot sooner than before, regardless of external opposition, in the renewed mind using God's brain.

In this expectant state, we can expect God to do what God does with visualization of His Image and Likeness becoming the stimulation to see us through the day.

The joy of the Lord is my strength and divine energy at work within the processes of the human brain.

This is the empowerment of the Spirit of Man within mankind, empowered to use the brain designed by God for Adam's body.

Once your mental thought process is expanded into, "I can do all things possible," there is open, unlimited favor; meaning whatever is going on, we will always have favor to make a negative issue in life turn out for the good as we renew our thought process to that of Christ Jesus.

This is due to the internal shift of "Joy of the Lord" as our strength, which is the strength we need to carry on and keep it moving, as we deal with the issues of life.

The stimulus is something that incites us to thought, feeling, or action, and the Holy Spirit is like an accelerator to our stimulus, working as a quickening power delivered from our divine intuition of what Love is, and what Love is not.

The quickening power delivers our mind from exceptions, and we can recall redemptive feelings, prompting us into redemptive action.

The individual who feels uncompromisingly favored and loved by God is likely to now move by his spiritual understanding versus his old, natural man understanding.

This redemptive reflection in the human brain that God designed for His Purpose and Glory directly results from the spiritual mind being opened to the unlimited Power of God's Love, Grace, and Mercy.

Now that should make you smile! The challenge is remembering the times we made the right choices, and the right thing happened as a result of that right choice, with an understanding that God made this thing simple, based on how to love Him.

During this race, we must overcome our Adamic nature, and society's antagonistic influence.

The mental neural pathways and thought processes of a wretch like me, and the variable of words, labels, and phrases that I used to visualize when I heard them, became the story I was telling myself.

The mental weight that I put on those words, labels, and phrases was because of the subconscious behaviors that I was taught to allow to be stimulated by opportunities of increase, and subconscious behaviors being stimulated by opportunities to prevent a decrease.

We can understand visualization as the same as "walking by faith, and not by sight." It was designed to produce a vivid internal road map for the human brain to focus on, directing the energies to produce the complex equations to be used for creativity, used by the body for physical manifestation.

It makes no difference what man might name this function in the human brain; it does not change God's original intent for its designed usage in the body that He made for His Glory and Purpose.

When a mental decision is made, a mental visual image is established of what's going to happen next --- because there is always a "What's next".

It doesn't matter how you are being treated, because you already know how you are going to respond to it.

You already know that your response will be one like a WWJD

(What Would Jesus Do?)

If you already know you are going to show self-love or love to someone you don't know, that can take the fear out of many things regarding life's issues because you already know what you are going to do: *love on them.*

Visualization is for mentally seeing God's love for me, and painting vivid pictures in my neural pathways, to create and manifest in my body His Likeness and Image for the world to see.

Here is the fun part. Close your eyes and visualize each of these.

"Come unto me all ye that are heavy laden, and I will give you rest."

Visualize that the Power within me is greater than what is in the world.

Visualize--- I am the head and not the tail.

Visualize the joy that's coming in the morning despite my current state of being.

Visualize that God is good All the time.

We just need to fill in the inputs of information to be visualized and put through the visualization that is marketed in society to help us be the best version of ourselves that we can be, especially in our decision-making process.

This created set of data comes from memories of events and issues that we have dealt with in the past and what stories we've told ourselves with regards to our perception of what happened in those episodes.

We now insert a logic-type: The logic of man and The Creator vs The logic of the Creator and man.

I will insert the Word of God out of my spiritual desire to love Him, or I will do something less than inserting the Word of God, which now can be computed as a zero in our logic gates.

This is exactly how He saved a wretch like me, once I used my free will to tell myself a different story about what is good for the body, and what is not good for the body.

After creating the perception based on my experience in choosing to lean not to my own understanding, I began to tell myself stories that I had not told myself before, and these became my forethoughts.

Yes, forethought is at the plate now, ready to take a swing at predicting what's about to happen for our future selves.

Can you afford the cost? Can you afford not to know about the cost?

The problem is, these complex cognitive thoughts take two seconds to happen, but our subconscious behavior can take less than a millisecond to manifest a thought and bypass into an action by a member of your body.

These are the subconscious behaviors, which do not know they are subconscious behaviors, that we have been doing subconsciously wrong for so long.

These are the bad habits that become part of our long-term memory, with a hair-trigger, easily stimulated release of the potential energy that is ready to take care of the job, allowing the brain to get back to running the other 10 billion neurons of the body.

This is the lack of knowledge that my people perish from. Tell me where this statement is wrong.

However, the truth is that "subconscious" behaviors are simply behaviors that have been thought of so often that the speed link of the neural pathway has become ingrained as automatic truth.

When people pop off at the mouth, they are saying ingrained subconscious "thought behaviors" that pass through the conscious thought process.

We know that when this happens, we say what we say, or do what we do, only to consciously say later, "I wish I had not said or done that earlier."

The good part is God's design of the human brain and its capability to continually rewire itself by adding and removing synaptic connections.

This is how we "learn" and create what eventually becomes a renewed belief system.

"Belief is by choice, and once you choose to believe something, especially about yourself, you will ignore all incoming information.
This is one way we limit what we are supposed to know, thereby limiting what we are supposed to be doing.
To be successful in life, we will ignore some information that will change what we do and how we do it." -A.R. Bernard.

I have found out that when you don't know the whole truth, it means you will believe a lie.

When you have knowledge of something, but not the whole truth, or better put, when you are taught something that is not true, and you believe it, you are believing a lie.

Take, for example, we all have the knowledge that the sun rises in the east and sets in the west. But if you were turned around as a child, and society taught you that west was east and east was west, and you still choose to believe it---you are choosing to believe a lie.

It does not change what the real truth is. You will react and respond to what you believe and receive the consequences, good or bad, from those actions.

Like, "I believe I can rob this store and get away with it." But the real truth is the store has a video surveillance camera connected to the police station.

But understand, *for real-for real,* the deep trick of believing a lie.

It will keep you from acting and responding to the real truth and receiving the benefits from those actions.

The brain continually rewires itself by adding and removing connections to neural pathways in order to add new things and remove old, putting off the old and putting on the new.

Does it make sense to look at spirituality from God's point of view, from the brain He created for His Glory?

It makes sense in my human brain that God's Breath into mankind would include the DNA to form a section in the grey matter of man's brain, that rightly divides His Word. It makes good sense that God would have made a way for man to understand how to use his gift of free will.

I did not learn this in church, but from the *"Lessons from the School of Hard Knocks"* and how I got over what society had shown and taught me.

I learned in my brain how Grace and Mercy changed my thought process to understand that God is Love, and I Am made in His Likeness and Image.

When you have a choice to make, it means you have a decision to make, which requires your brain.

It is my conclusion that there is something inside mankind's brain that came along with the Breath that God breathed into Adam's nostrils.

God made this thing simple, and it is man who complicates it by the choices he chooses to input into his brain, equate with his mind, and manifest in his body for the world to see.

God is Love (the input for the brain), the energy source of all Creation (has the equations for the mind) that will physically manifest in the body.

Love is the energy force that has the equations to correct the variables from Adam's mental decision to be selfish.

We come into this world with an innate desire to love, and we are taught how to hate, with negative words, labels, and phrases that make the human mind visualize something that is not true.

Zeros and Ones reverse-engineers the thoughts that do not benefit your future self as you deal with the issues of life.

The journey, it seems, is about using the right variables of words, labels, and phrases, which make your human mind visualize how to use the Energy Source for your future self. When we insert God is Love, the Force that can move all other forces, we are acknowledging that Love is the greatest and most powerful thing in the universe.

The Dunamis Power of God's Word inserted into the thought process manifests actions and beliefs that point towards Love.

What's going on in our brains when we say, "Greater is He that is in me, than he that is in the world"?

What's going on in our brains when we say, "I can do all things through Him who strengthens me"?

What's going on in our brains when we say that God gave us a spirit not of fear, but of power, love, and self-control?

What stimuli are produced in our brains, and more importantly, what is going on in the brain when they are not said?

It is critically important that we start a conversation with what is going on in the brain that God designed for His Purpose and Glory.

We receive information from our five senses, which are in contact with everything outside of our own bodies; and we send out information through those senses to everything outside of our own bodies.

When dealing with the issues of life, it is our perception of that issue that we will label and encode into our brain.

Societal norms are set in place to bypass conscious thought, and this is transferred in the form of teachings from an adult to a child using words, labels, and phrases that make the child's mind get a visual image of what the taught word, label, or phrase assigned means.

This cycle is not new news. However, what is new, is the concept of trying a different way to heal and become one.

God's Brain, for God's Mind, for God's Body: the way He designed it to manifest His Love and Light.

Making a choice involves a preference for one course of action over another, requiring a sense of how choices will turn out.

It is free will and forethought based on the Logic of Zeros and Ones, which ultimately determines what value we put on the choices before us in the decision-making process.

To the Point

Word Knowledge: Perception
The result of receiving and mentally decoding and interpreting data from the five senses.

Word Knowledge: Encode
The processing of information from one system of communication to another. Encoding permits the downloaded information to be formulated into a construct to be kept within the brain and recalled later from long-term or short-term memory.

Word Knowledge: Memory The function of memory is to decipher, download, store, and recall information. Memories offer the capacity to comprehend and adapt from past experiences and relationships.

Word Knowledge: Working Memory Mentally archived information for prompt use (the thought process for what you know) to be aided by attaching to previously stored items already existent in an individual's hippocampus and the long-term memory.

Word Knowledge: Label When describing who you think you are, a label is what you mentally call yourself and the memories and information to which you affix yourself. We place labels on ourselves and others to define who we are.

Chapter Eleven: Eleventh Grade

Neuroplasticity and the Creator's Design To Get Over Adam's Mental Mistake.

How to Put off the old man and put on the new man.

LFTSOHKS CURRICULUM IS a course study focused on how to overcome the faulty third-party programming we received in our formative years.

If we do not fully understand how we think and why we think, we allow others to think for us.

The curriculum for life is to understand that the brain was designed with the ability to rewire and rewrite itself.

This is how we evolve and grow. The goal is to teach our youth to see life from a spiritual point of view.

This perspective is essential due to the deviations in societal norms that influence the experiences that shape neural pathways and impact the decision-making of our youth as they transition into adults.

The curriculum for life is composed of mathematics, literature, and biology.

There is a good chance that the words (literature) you say will cause me to have a physical reaction, (biology),

My physical reaction (biology) will cause you to do some mathematics and add the things up based on what you perceive from my reaction (mathematics) to determine the story (literature) that you will tell yourself.

From that story, you will produce (mathematics) chemicals in your brain (biology) that will influence your next physical actions (biology), which represent your final decision, (of mathematics) that will be manifested in a physical (biology) way, from your body for the world to see, and interpret the perceived meaning through known words, labels, and phrases (literature) that have been manifested in your physical,(biology) body and the mathematics of your mind.

The curriculum for life is composed of mathematics, literature, and biology.

A curriculum of words, labels, and phrases is designed to convey someone else's thoughts into a visual form that can be encoded into your brain.

Neuroplasticity and the Creator's Design To Get Over Adam's Mental Mistake.

We must understand the emotional literature in the language of words, and the biological effects that emotional responses to those words, labels, and phrases can have on the biological chemistry of the brain.

Subconscious thinking can slip past the conscious portion of the brain and communicate directly with the motor functions of the physical body.

That is the point where subconscious understanding bypasses conscious thought.

We don't think about breathing because we do it subconsciously. However, if we could not breathe for some reason, it would become a conscious thought.to remember to inhale a certain way through your nostrils for two seconds, exhale through your mouth for four seconds, puckering your lips in a circular motion while breathing out slowly to relieve an anxiety attack.

Many times, physical things happen that we are not conscious of, until we are facing unintended consequences of what we have already manifested in physical actions.

"I pulled the trigger before I knew it." "I laid hands on them," or, "I said what I said, before I knew it." (biology without math equals negative literature For your future self).

The Curriculum and the Derivative of Love

"Baby, you know I love you, and I promise I won't do it again"

This lie creates changes in the chemicals produced by a woman and can shut down the neural pathways that lead to a positive energy transference of uninterrupted love. There is a thin line between love and hate, and broken trust is right there at the thin line. When this occurs, the neural pathway that used to lead to love now lead to a new pathway that says ...

"You better sleep with one eye open."

Another example is feeling negative energy after a binge-eating episode caused by some disturbing news you got earlier that day. Here's the science behind it: binge-eating results in the release of dopamine that's responsible for feelings of reward and pleasure

The dopamine drop from binge-eating has already been certified in your belief system and is on the fast track of the path of least resistance that says after a disappointment, "You should, because you can," "Go ahead, indulge in half a gallon of butter pecan ice cream and all the dark chocolate you want." This is free will in full effect when you wish it wasn't.

The dopamine drop will temporarily help to balance out the negative drain on your happiness.

Because we are unaware of these subconscious behavioral thoughts, we make inexperienced decisions in "rightly dividing" the words that make up the story that we tell ourselves.

Neuroplasticity and the Creator's Design To Get Over Adam's Mental Mistake.

These inexperienced decisions create the stories that we recall from our memories before we make decisions, and lead to negative manifestations.

The goal of the curriculum for life is to know the end from the beginning.

Teaching the end from the beginning will produce a thought process that has more information about the Power of Love and reveals with more clarity and more often how emotional stress leads to negative energy, and negative production of the human body.

The parallel equations for robotics, when programmed into the robot, will bypass and shut down the old electrical circuits that produce negative production, at the same time, rewiring and reprogramming the robot to produce positive movements that produce positive production as designed by the thought process of the creator or the programmer.

The curriculum will create positive neural pathways that are hard to bypass in times of emotional stress, because the foresight being included in the mathematical equations is something that our present selves can include in the decision-making process so that the pathway to proceed down will be enlightened…with wisdom.# Who's the Author of Wisdom?

Let's all say it at one time: "God Almighty Himself, the Creator of All Heaven and Earth!"

By understanding that God is the Author of Wisdom you will be able reverse engineer the negative thought process of Adam.

God's Brain for Gods Mind for Gods Body – full circle back to the way it was designed to be used with the Wisdom of God, and not the thought process of Mankind, who was born in sin and shaped in inequity, by other human beings shaped in inequity {drop the mic.spiritmanspeaks.com}

As I am thinking that the thinking, that has brought me this far in life, has presented me with "problems that this thinking can't solve," then "I" thought, "That's because this thinking that I am 'thinking' is a derivative of the thinking I was designed to think with, programmed in me by others who were designed to think with the fallen thought process of Adam.

Division by Selfishness

Adam's thought process and the subconscious behavior of "the path of least resistance" is one of the places where the twisted gets twisted at, because its corrupt.
> how you go about multiplying and increasing,
> how you go about replenishing, and
> how you go about subduing:

all now divisible by selfishness, greed, and the desire to have power (Sovereign Authority) over other human beings.

The path of least resistance to do this is to have a lot of "something" that stimulates the thought process of people.

Neuroplasticity and the Creator's Design To Get Over Adam's Mental Mistake.

Number One is Money. That will get you the perceived power that was connected to Mankind's Oneness with God that he had before the mental miscue. Now all that has been taught in the mind of mankind is a derivative in our thought process today. #The love of money.

In my opinion, the Basic Instructions Before Leaving Earth is the curriculum for the mind of those who will have your free will choose choices that God the Creator of All Heaven and Earth chose for YOU to choose Him.

By using your Free Will to choose a "one", not a "zero" to shut down neural pathways that do not produce more neural pathways that produce more Positive energy, positive production towards
    0 times 0 equals 0,
    0 times 1 equals 0,
    1 times 1 equals 1
    Positive energy times positive energy equals positive energy
#One Love #His Image #HisLikeness

The path of least resistance and programming through the story you tell yourself becomes a filter created for your TRAS-R ; allowing the renewed mind to operate as The Designer designed to solve the problems that the natural mind of man cannot solve.

Case in point: "I can do all things with the thought process of Christ Jesus" bypasses what society has taught me to think that I can or cannot do. So, when I insert it into my thought process, as I get to the obstacles in life that are in my way, the creativity of The Power that's inside my belief system accelerates my progress towards proper exercise of my Free Will to have Freedome to exercise Freedom vs. the belief that I am something less than a Child Of God; because, over time, I have seen how much better it is to think with my renewed mind compared to thinking with the mind that was born in sin and shaped by other people also born in sin.

The antagonist in the book of your life, can only change the color and style of his outer garments: but internally he remains the same antagonist. #Don't get it twisted!

This is the one trick (subtle deception) which since the Garden continues to trip us the human mind.
#Surely God did not say.
#nothing new under the sun.

The Fear of Missing Out (FOMO)

The Original FOMO is really the spirit of Man's fear of missing out on heaven. That fear is really the stimulation of trying to prevent the decrease of "not being in Oneness with God."

True or False: The above is subconscious behavior being brought up for a conscious conversation.

The Curriculum of the Five-Year-Old

If you have ever left your cell phone with a five-year-old, you probably noticed that their intuitive, rapidly changing mind can

Neuroplasticity and the Creator's Design To Get Over Adam's Mental Mistake.

figure out how to do things on your phone that you did not even know were possible.

This raises the bar on your own thought process about the technological abilities of children.

When you see another young kid do the same thing with a phone, it strengthens that particular neural process—these young kids really know what they are doing!

This causes you to watch your phone a little closer the next time you hand it to your five-year-old nephew.

While we were building that neural pathway, something else happened: the brain concluded that the younger generation understands technology, so asking them to help us fix phones, computers, and TVs now makes sense. (The path of least resistance in full effect)

Those over fifty-five might remember wondering "Why in the world would your grandparent call you all way from another room for you to come change the channel on the television.

the human remote control, before remote controls became affordable and available for consumers.

The path of least resistance: being the human remote control before remote controls became affordable and available for consumers.

The path of least resistance rewrites the script to "Use the resources of younger minds and take the easy way out of technology training", in order to let the brain, get back to running everything else in the most efficient manner.

This neural pathway is included in our plan of accomplishing a goal. The LFTSOHKS is the evolution of reprogramming our minds to develop and use neural pathways that are based in love, because in that state is where we are our best selves. We are born with the innate ability to love, but we are taught how to hate.

There is a new understanding when we incorporate this new knowledge into our thought process. It changes the zeros and ones of a programmed neuronal firing sequence that turns a thought into an action.

These newly implemented subscripts of stimuli will now cause the physical body to produce different actions.

It's my conclusion that this book will produce a stimulus in the spiritual mind of anyone who reads it for the purpose of creating an impact on subconscious thought.

"...Greater is He that is in you than he that is in the world."
When you break that down and begin to tell yourself that story, with words that stimulate the innate mind that God designed, you will create neural pathways that will plug in to the Greatness within.

"...Greater is He that is in you than he that is in the world." has a whole new meaning now, being filtered through the TRAS-R.

Neuroplasticity and the Creator's Design To Get Over Adam's Mental Mistake.

The curriculum for life will teach you how to acquire knowledge and wisdom, and filtered understanding in a way that will create neural pathways resulting in the manifestation of positive outcomes.

> *"When you become the master of your mind,
> you are master of everything."*
> *-Swami Satchidananda*

The word 'curriculum' has roots in and derives from the Latin word for "race-course." As the meaning continued to evolve, we see its use more recently dating to 1902, when the word loosely translated to "the course of one's life."

This view explained the curriculum as the course of deeds and learned experiences through which children transition into adults.

As this transition occurs, an individual learns from the past and uses those lessons in becoming his most authentic self. #His Likeness His Image

The curriculum for life is the process of navigating and interpreting those past experiences for the betterment of your current and future situation.

Again, the curriculum for life is to understand that the brain was designed with the ability to rewire and rewrite itself.

Reprogramming yourself from the past starts with learning, analyzing, and redirecting habits to better your own human experience.

The curriculum for life is growing the neural pathways that produce verbal and physical actions that point to love and positivity.

Implementing this curriculum into your life requires a merger of spiritual understanding with regards to the science of the body God made for His Glory and Purpose.

Humanity has the opportunity to transcend oppression and strongholds and reignite the brain's ability to visualize the image of love to manifest action.

This is how love is manifested for the world to see today.

The curriculum is about learning to see life from the Creator's point of view and equipping our young boys and girls with a spiritual perspective that will result in their best future selves.

It's my conclusion, that when "those who will" read the Words from this book, it will produce a stimulus in the Spirit of Man that will understand this on a subconscious level, and now manifest in the members of the body: to do it.

How? With learning materials such as *Lessons From the School of Hard Knocks*, assessment guides, teacher guides, and textbooks that will add value to the thought process of mankind. Serious business, but simple math; Zeros and Ones *Lessons from the School of Hard Knocks* helps mankind choose the One more clearly and more often.

This is the part where we all step off the ledge and take that leap of faith, which is the ability to change your belief system to

Neuroplasticity and the Creator's Design To Get Over Adam's Mental Mistake.

point to Love, using the Dunamis Word of God as the stimulus to change the thought process of the Spirit of Man to acquire knowledge to filter out the things that are not good for your future self. #Foresight to see the end from the beginning.

The curriculum for life uses God's brain, God's Mind to rightly divide the Word to be manifested for God's Body and the perception of those norms that will create the story that you tell yourself.

The curriculum for life is the ability for the brain to imagine the image of God/Love, and manifest that Image and Likeness for the world to see.

The curriculum for life is used to foresee and visualize what is positive for your future self even though your present self will not see what is positive, until the outcome happens.

The curriculum for life, is knowing that when you have a choice to make, it means you have a decision to make, which requires that you use your brain.

The curriculum for life is having a better understanding of the comprised neurons, connected by dendrites created in the brain, Based on our habits and behaviors. #for real for real

To the Point: The rEVOLution has begun: The Evolution of Love.

Chapter Twelve: Twelfth Grade

Stop Tripping in Your Flesh and Walk in Your Spirit

It is a prison ministry sermon that was originally written on paper and delivered at Muskegon Correctional Facilities, 2001 written in a Homiletic preaching style, over twenty years ago.

This sermon was the story/words labels and phrases I had told myself and was continually telling myself that helped me to stop tripping in my society taught mental thought processes, which produced the tripped-out issues in my life. This story completely shut down some old Brooklyn Hustle neural pathways because those words labels, and phrases had negative false variables that did not stimulate past my TRAS-R as a benefit of positivity for my future self.

Lessons From the School of Hard Knocks was needed 2000 years ago. It was needed 20 years ago, and now without a doubt, it is needed today to help the thought process of humanity see more clearly, and more often what produces confusion and where it comes from.

There is nothing new under the sun.

>PTL, Honor, turn to your neighbor, I'm glad you are here, Jesus wherver 2 or 3 are gathered in my name I will be in the midst. Tell your neighbor I need you and you need me...because you are here & I'm here...Jesus is here also (if you know like I know...If you have the knowledge understanding & wisdom -of who Dwells In the Midst of praise ...(of what happens...When the praises go up ...etc. will you put those hands together & praise. our Lord & savior Christ Jesus....When you realize, that Jesus is here Right now...all you have to do is tell him what you want)

>You have to understand. What this service is for ...The Great invitation ...Jesus said ... Come Unto Me ...All Ye that labor ...And are Heavy Laden. And I will Give you rest}

Give you peace of mind if you have been laboring /working on what's the deal if your mind is heavy laden with what's going on with the who, the what, the why and the how etc.}

- Before this service is over... His word will give you instruction on how to recognize And Receive His Will for your life. (etc. I want to be in the will of God)I need you and you need me.

(the spirit man and how it has to overcome the natural man ...on a continuous basis... in its ability to do the right thing at the right time...

Understand (Rom12.5... We have many members- but one body. In Christ)-{The body of Jesus was God Manifested in the flesh...(his thoughts, his actions,his character) where reveled in that body(areyouwithme) we are here today gathered in that Name...Jesus... to receive Knowledge and understanding of the mental Process ...(the series of changes, our minds have to go through ... of how to attain and retain the manifestation (we can do right thing today and wrong tomorrow)...of his thoughts, his actions ... his character... into the members of our own bodies—and many members of his body ... on earth... The church... Which consists of people... From various backgrounds ...who are in various stages, & levels. Of accepting Jesus Christ... For who he is & What he can do ...(and the reason for the various levels... is the various levels of trust ... of letting go ... and letting God... letting Go...Of what society has taught us and shown us... And letting go of the sinful state ... we were born into. And trusting the word of God (pause)

So, the question becomes... What's has taken us... so long to get on the same page – as Jesus ... and when I ask myself that question...(What took me so long to get with the program ... to get on the same page of thought as Jesus... and the honest ans... would be... What for...Or what's in it for me...Isn't it true that Society has taught us to ask ourselves...

That all motivating question "what's in it for me "does that question sound

Familiar ... What Motivates us...to do the things that we do...(it's a Psychological fact ... the # 1 motivator –is personal gain...(and all the honest people said (amen)

What's in it for me ... why do I need to act in a manner that represents Christ ...

But little did I know (and I mean little did I know) the honest answer would be ... I did not know,what I know now ... I did not have the understanding... Of what I understand now ... my mind could not COMPREHENDED... The goodness of Jesus... And the benefits of his promises ... my mind could not conceive... What my eyes. Could not see...

I came today to share that Gods Word is the SAME, YESTERDAY, TODAY and will be tomorrow. and his promises (the ans to what's in it for me) are the same yesterday today & will be tomorrow...

Just as he Promised Israel... The land of milk and honey – he has the same promises to us today...

Turn with me to the book of Joshua... Chap 1 vs 1-9 and we will be REFERANCING THE book of Exodus 20 & 32(read <pray >)topic crossing Jordan subtopic... Stop Tripping in Your Flesh. And Walk in the Spirit

From Bible class, and Sunday school... We know the people of Israel... Have been wandering in the wilderness... For forty years) and in our text we see they are about... to cross... The Jordan River (set stage which is raging high tide)... But if we go back 40 yrs. to the exodus at chap 32 where the scene is (read 10 commandments)

Moses delayed coming down from the mountain top..., and they did not believe he was coming back ...

> understand this about belief, Belief is by choice and once we choose to belief something...Especially about yourself ...you will ignore all incoming information (2x)

>This is one of the ways we limit what we are supposed to know... Thereby limiting ...what we are supposed to be doing... To be successful in life,

Society will try to teach us that we are poor and no good because of our economical back GROUND,

Where we come from or even the pigmentation of our SKIN.

So, if...we believe we are supposed to be poor and no good because of ECONOMICAL BACKROUND, where we come from... or the color of our skin ... we will ignore some INFORMATIOM THAT will change what we do and how we do it ...

Orrr-rum This type of societal influence is presented to us in our teen years...

Pause" some of us of did not believe the presentation from society ...but a sad truth is...Some of us did...that's one reason that the drug hustling life and the drug abusing life... never runs out of participants... which leads to one of the biggest business in America ...the prison system...

> I've found out "when you don't know the whole truth... that it means you will believe a lie (2x) "when you have knowledge of something ...but not the whole truth... or better put when you are taught something... That is not true. And believe it (you are believing a lie) a good ex would be

>(We all have the knowledge that the sun rises in the east and sets in the west......But if you where turned around backwards as a child and society taught you that west was east and east was west ... and you still choose to believe it... You are believing a lie ...

>It does not change <u>what the real truth is</u>... You will react and respond. to what you believe... and ... Receive the consequences (good or bad) from those actions...(I believe I can rob this store and get a way with it ...but the real truth is the store had a video surveillance camera hooked up to the police station)

But understand the deep trick of believing a lie... <u>it will keep you</u> from acting and responding to the real truth and receiving the benefits from those actions (aywm)

So... Here in our text ... we have the Israelites in exodus choosing to believe... that Moses ... Was not coming back and they ignore...Some very important instructions and information from God (read exodus20)

God lets us know what time it is... he is very to the point of what is good and what is bad ... Understand there is no faulty gray area when it comes to the word of God...It is man who has the faulty gray area... in his interpretation ...and... UNDERSTANDING WHAT is good and what is bad for the body of Christ (2x) (pvb 4&7)

And when we look at ex chap 32,2-9 read so the mentality is lets take our gold...lets make us a calf looking god ...lets make an alter... Bow down and worship to the calf looking god for bringing us out of captivity. And then tomorrow lets play and celebrate and have a party... Down the block at the club or at the elks lodge with our god the calf...

...we can plainly see... this generation...of Godly people...Where straight up ...Tripping, that day... tripping in there flesh ...and not walking ...in the spirit of god ...that had led them out of captivity from Egypt...(oh how soon they forgot what the lord had delivered them from)

And when we ignore and forget that Jesus died on the ole rugged cross...And became our light to led us out of captivity from Sin... ...we too will experience episodes of tripping in the flesh –and not walking in the Spirit of God...with the members of our bodies (hands feet &mouth)

This leads to a rhetorical question? Have you ever said anything...You regretted saying ...have you ever done anything you regretted doing ... have you ever looked back at some of the things you use to do with astonishment etc....and all the Honest people said Amen)

The reality of today... is we have to get to a point to where we are Honest enough with ourselves to where we can plainly see...when we are tripping in our flesh –and not walking in the spirit of God... pause (something to chew on) if you can see and understand the problem... You should also be able to see and understand. The reward of overcoming the problem (2xetc I got have a problem with cig, alcohol, and all the drugs the reward of overcoming those problems is better health, I got a problem with managing the incoming and outgoing of my finances,The reward is better wealth.

back to our text... Joshua and the Israelites had been in the wilderness 40 yrs. their leader Moses just died, now they come to the raging JR they must cross...without a boat or a ship or raft ...pause And I believe they looked at this situation and at each other and said no big deal...((soup Campbell) this is no problem ...they had been through so much) and over the course of time 40 yrs. ...their thought process had been changed...No longer where they afraid of what God...could not do... they had mental activity with a the series of changes in...that their minds went trough... when they saw this raging river and... the final result... the conclusive thought was... that God is a Faithful God) (that has been with us through it all and has never failed us and and all we have to do is remain obedient to his instructions and he will supply all our needs.

SO the application in today's reality is when we are faced with major Obstacles/Jordan rivers in our lives... if we can just have the mental thought process... That God is faithful God. if we can just remember where he has brought us from) if we can just remember the last time we where in a world ending situation up the creek with out a paddle, and the LORD made a way... out of no way and proved that he was a on time God (oh yes he is) ... and there is no good thing that he would withhold from those who walk upright before him ... turn to your neighbor and tell them to be strong and have good courage... and now is not the time to give up... or give in

...I heard somebody say I know to much about him (turn mike) for you to make me doubt him ...ohh and I heard somebody say weeping may endure for the night (turn mike) but Jooyyy will come in the morning...(I love when the anointed hooping preacher says these things...and as I understand it the inner spirit man gets strength from the power that's in these word and it sends a spiritual emotional charge to the body...that will make you shout... That will make you dance...And in my church do laps around the sanctuary) ...hallelujah) I love the spirit man inside of me ...simply because it will make me do right... when I want to do wrong ... AS the inner spirit man grows in strength, it will help us walk in the spirit of the lord and adamantly remind us not to trip over our flesh ...and the honest people said...Amen

(Verse 3and 4...Then God told Joshua...How large... The land would be ... the Israelites where to have} (If) they should show them self's worthy (read) ... in this verse it gives the example and explains, that there are some pretty big ... things in store (for those ...who are worthy) The dictionary tells us worthy means –of equal value in the eye of the beholder ...(the way I see it and the way you see it... it may not be worth it to you but its worth it to me ...it makes sense to me but it may not make sense to, you may not see the worth or equal value in keeping your mouth closed... in certain situations...as Christians who are diligently seeking the deep things of God we should know the worth and value of not letting the dev. Get a foothold... the saints of God should see the V&W of renewing our minds with that of Christ ...the way we think...The way we understand to calculate 1+1 =2, to get to the conclusion. Of what is good for the body of Christ and what is bad for the Body of Christ

Our text tells us three times to be strong and have good courage, which is the ability to conquer fear, and overcome things that intimidate us (2x)... we need to conquer the fear that God Can't do what he said he would do...The instructions and information from God is to be strong and have courage... to strengthen ourselves...

And I'm glad you asked the question... how do we do that? The answer my friends is it with his Word (scrip)... Jesus told the dev it is written... so when we come to the banks of the raging Jordan rivers in your life) you will not be intimidated ... you will not be afraid to choose the word of God...if you know what the word of God says...But ""if you don't know what the word of Gods says... you wont know what the word of God can do... (let me say that again for the honest people 2X) and if don't know what the word of God Can do ... we will lean...to our own understanding ...where the word God says lean not to your own understanding... we will acknowledge what society has taught where the word say to acknowledge him in all thy ways ... and our direction will be down a wrong path. ...where the word says he shall direct our path ...this application of prv 2&7??/ is played out every time...it time to cross the Jordan rivers in or lives...

Understand when we look back upon our life.

Reality tells us we are where we are because of choices that we have chose (2x)

But That's nothing new God gave man the ability to choose(the reality is, we are we are because of choses (that we chose)and when we believe the truth that God has already made an way of escape, we will react and respond and receive the benefits from those actions

But he also gave us good instructions Pvbs 42,4&2 ?tells us...for good instruction is what I certainly shall give you THE REASON FOR THESE INSTRCUTIONS MY Friend IS SIMPLE...God in his infinite wisdom...in his knowing the end from the beginning ...knew we would need a way of escape …from the consequences of the choices that we… choose that did not line up with his instructions or in other words he has already made a way of escape …he has already made a way of escape through his WORD he has already made a way of escape through his son Jesus Christ … John 1&1 the word was made flesh, instructions and information that when followed will lead us from the captivity of society has taught and shown us and… instructions and information that when followed will lead us from the our the sinful nature we was born into.

But if we choose to ignore the information we will continue to trip over our fleshly desires (areyouwithme)

As we close

Understand we are put into a position to make choices between right and wrong, good and evil to trip in our flesh or walk in the spirit of God we are put in a position to make choices and show if we are worthy…of the promises of God…
in a position to show equal value in our praying,
a position to show equal value in our praising
a position to show equal value of faithfulness an equal value in our fasting and tithing…(aywm) can I hear from the honest people …
and somebody here today is in a position to show if you are worthy of the promises of God…

Abraham was put in a position to choose and show it with his son Isaac...

Job was put in a position to show to choose and show,

Adam and Eve was put in a position to choose and show...

Peter was put in a poison to choose when the Lord asked him 3 times

Peter do you love me more than these (3x) And Peter and you know I love you Lord...And...

That's how you cross over to other side of the Jordan rivers in today's reality by loving God...More than these ...
more than these things of the world
by loving God more than the things that are seen In the world.(t.v. mtv... society has shown you ...
by understanding that things of this world are temporary and the things of God are everlasting...
and that by loving God ...
your anger can no last...
BLG your hatred cannot last...
BLG your Jealousy cannot last
BLG you wont even be able to Gossip as much...and all the Honest people said AAMEN...
...help us Holy Spirit...I'm not going to talk about my brother or sister while they are tripping in there

flesh. I am going to pray that your grace and mercy be sufficient until they are able to walk in your spirit.

I am going to pray what you did for me lord while I was in my mess...You will do for them... they might have done me wrong but I love you lord and
... I am not going to trip in my own flesh but I am going to walk in your loving spirit ...
because its worth it...To see you in glory...
Because it worth it to hear you say well done...my good and faithful servant. And all the honest people said?...Aaaamen.

>>In my closing this text is about Crossing the Jordan river... a Wide of space from this side to the other side. I believe the widest space we have to cross in today's reality... is the wide space of doubt and confusion and in our mentality in our thought process...

We have a wide space of doubt and confusion from the natural, to the spiritual ...a wide space of looking at life from societies point of view
to looking at life from Gods point of view ...
it seems that some are hesitant or have an unwillingness to crossover living for the world... To living for Jesus...

And that space of doubt and confusion...

Has a direct affect on mans free will and will influence what he sees...

What he feels and what he believes. what he chooses and ultimately his place of everlasting life, (hell wouldn't be so bad if we only had to stay there 10,00yrs)...

That's how some people see what they want to see and believe what they want to believe ...

But as for me and my free will...I might have been turned around as a child and taught east was west and west was east... bur I thank God He shined his light on me... and I was able to learn to accept the real truth ...that east was east, and west was west ...

I was able to accept the truth it is better to please God than to please man...

I was able accept the truth and believe that God Can do all everything and I can do all thing through Christ who strengthens me, so when I come to banks of the Jordan River ...

The width and the space of doubt and confusion in my thought process is bridged by my belief that Jesus died on the Cross went down below Got the Keys... came back up and proclaimed ALL power is in my hands...
now I got the power to cross the Jordan rivers... Has anybody else in here got the power to cross over to the other side ... can I get some honest people to

stand up with me, can I get some honest people to cross over with to the other side

I came to tell the story that you already know and that's, that the adversary dose not want you crossings over to the other side to land of milk and honey and to experience Life in Christ ... Jesus

Understand The Same God That told Joshua., I will have the same relationship that he had with Mosses, he would have the same relationship with us, IF you just Follow Jesus you have good success ...in crossing the Jordan rivers

Time TO stop tripping Time To start understanding ---what you did not understand before

Thoughts leads to actions...as we learn the process manifesting Gods words into our actions...the benefits...will be...
(read) Gal 5&16

1st Corinthians tells us the spiritual man searches all things... yea...the deep things of God... and glory to God I already know this is not no Kindergarten church...
this is and we are diligently seeking the deep things of God and all the honest people said... Amen ...

This is the deepness, this generation, of Godly people must understand..." what is it worth " as respond to the Issues of life...

'Is it worth saying something back. is it worth retaliating... and going" (don't go there) ... what value...

Is it me in the long run ...(if you can see & understand the problem...(the Jordan river)...
you should also be able to see and understand... The reward of overcoming the problem... (Crossing over... that Jordan river)
(read from bible verse 12) then Joshua gave orders, to his officers to go through, the camp and tell all the people to prepare food for the journey... for in three days... we shall pass...over... the Jordan river... and go into the land, the lord has promised...

(I believe this word ...into today's reality... letting us know we need to be prepared with this spiritual food... this manna to cross over mentally, and use the brain and mind for the members of our body to show the love of Christ to others Don't be scared to love, Don't be scared to trust God

(Thank you Grandfather)

ONE THING I KNOW FOR SURE IS THAT IT IS VERY DIFFICULT FOR THE ADVERSARY AND THE DEVIL TO INTIMIDATE ME.

Stop Tripping in Your Flesh and Walk in Your Spirit

IT IS HARD TO INTIMIDATE SOMEONE WHO IS STANDING ON THE TRUTH.

NOW THAT YOU KNOW SOME THINGS FOR SURE, YOU CAN STOP BEING SCARED OF THE DEVIL.

WHEN YOU KNOW SOME THINGS FOR SURE, YOU CAN MOVE FORWARD.

WHEN YOU KNOW SOME THINGS FOR SURE, YOU CAN GO TO THE NEXT LEVEL IN CHRIST JESUS.

WHAT I KNOW FOR SURE IS THAT IT IS BETTER THAN THE LAST LEVEL.

MORE PEACE OF MIND, LESS STRESS, AND LESS WORRY.

ACT LIKE YOU KNOW.

PEACE WITH GOD, PEACE WITH WE, PEACE WITH OTHERS, AND KEEPING THE PEACE.

-END- (Alter Call, Pray Out, "Thank you, Lord")

Post Chapter Twelve- 2022 Edition

Crossing Jordan, Stop Tripping in your flesh and walk in your Spirit

Praise the Lord ... Grace, and Peace

Prayer --- Lord bless my eyes to see, my ears to hear, and most of all bless my heart to receive Your Word in truth.

We are gathered here in the name of Jesus to receive knowledge and understanding of the mental process and series of changes our minds have to go through to attain and retain the manifestations of God's character, God's actions, and God's thoughts.

"For as the body is one, and hath many members, and all the members of that one body, being many, are one body, so also is Christ."

The body of Jesus was God manifested in the flesh.

This is a service for "the great invitation" because God said, "Come unto me, all ye that labor and are heavy laden, and I will give you rest" and peace of mind.

Before this service is over, one everyone should recognize and receive God's will for your future self, for your life; letting go of what society has taught us and accepting Jesus for who He is and what He can do; letting go and letting God; trusting the Word of God and all His promises.

I came today to share that God's Word is the SAME! Yesterday, today, and will be tomorrow; and so are His promises. Just as He promised Israel the land of milk and honey – He has the same promises to us today.

Understand this about belief--- belief is by choice, and once you choose to believe something, especially about yourself, you will ignore all incoming information. (Stop and read that again)

This is one of the ways we limit what we are supposed to know; thereby limiting what we are supposed to be doing. To be successful in life, society has taught us to ask ourselves, "What's in it for me?" And to be candid, this is straight from the selfish thought process of Adam's mental decision of bone of my bone, flesh of my flesh.

God lets us know what time it is, and He is very to the point of what is good for the body and what is bad for the body.

Understand there is no faulty, gray area when it comes to the Word of GOD. It is man who has the faulty gray area.

As we deal with the issues of life, we ignore and forget that Jesus died on that rugged cross, and lose focused thought with regards to the Power of Love.

Therefore, we are not on the same page or accord with God and His Word. The body of Christ, as I understand it, it is an energy force that motivates the human flesh to manifest in its actions as an image of their interpretations of what would Jesus' manifestations look like—also known as WWJD. With the end game being manifestations of our energies to produce more, we need to increase and multiply more energies that replicate feelings of Love. When we ignore and forget that Jesus died on the rugged cross and became our Light to lead us out of captivity from sin, we too will experience episodes of tripping in the flesh.

From Bible class and Sunday school, we know the people of Israel have been wandering in the wilderness for forty years, and in our text, we see they are about to cross The Jordan River. The stage is set in a raging high tide. Joshua and the Israelites had been in the wilderness for 40 years. Their leader Moses just died, and now they come to the raging Jordan River.

They must cross, without a boat, or a ship, or raft. I believe they looked at this situation, and at each other and said, "No big deal. This is no problem." They had been through so much, and over the course of 40 years, their thought process had been changed. No longer were they afraid of what God could or could not do.

They saw this raging Jordan River and had *mental activity with a series of changes* in their thought process.

The conclusive thought was that God is a faithful God Who has been with us through it all and has never failed us. All we have to do is remain obedient to Hs instructions, and He will supply all of our needs.

So, the application in today's reality is when we are faced with major obstacles or Jordan rivers in our lives. If we can just have the mental thought process that, "God is a faithful God." If we can just remember where He has brought us from, and if we can be strong and have the courage to trust our memory of the last time, we were in a world-ending situation, up the creek without a paddle, and the LORD made a way out of no way.

He proved that He was an on-time God! (Oh yes, He is!) And there is no good thing that He would withhold from those who walk upright before Him.

Turn to your neighbor and tell them to "be strong and have good courage", and now is not the time to give up or give in.

"Be strong and have good courage" in knowing that you know too much about Him, for anybody to make you doubt Him.

"Be strong and have good courage" that as you read and DO the teachings of His Word, your mind is being transformed as that of Jesus Christ, and that Spirit Man gets strength to control these earthly vessels, so that it has the Power to do so HIS Will.

Paul demonstrates the complication of our dual nature in Romans 7:15-25. "Now if I do what I will not to do, it is no longer I who do it, but sin that dwells within me." *(Romans*

7.20) I find then a law, that evil is present with me, the one who wills to do good, for I delight in the law of God according to the inward man.

I love the Spirit Man inside of me, simply because it helps to keep me focused on doing right when I want to do wrong, as the inner Spirit Man grows in strength, it will help us to walk in the Spirit of the Lord and adamantly remind us not to trip in our flesh.

And the honest people said, "Amen."

...Then God told Joshua how large the land would be that the Israelites were to have, if they should show themselves worthy. In this verse, the example is given and explained that there are some pretty big things in store, for those who are worthy. The dictionary tells us worthy means of equal value in the eye of the beholder. The way I see it, and the way you see it, it may not be worth it to you, but it's worth it to me. It makes sense to me, but it may not make sense to you.

You may not see the worth or equal value in keeping your mouth closed in certain situations, as Christians who are diligently seeking the deep things of God. We should know the worth and value of not letting the devil get a foothold.

The Saints of God should see the value and worth of renewing our minds with that of Christ, the way we think. The way we understand to calculate 1+1 =2 to get to the conclusion of what is good for the body of Christ and what is bad for the Body of Christ.

Our text tells us three times to "be strong and have good courage," which is the ability to conquer fear and overcome things that intimidate us. We need to conquer the fear that God can't do what He said he would do. The instructions and information from God are to "be strong and have the courage" to strengthen ourselves.

The question is, "How do we do that?" The answer, my friends, is with God's Word of the Good News. Jesus told the devil "It is written..."

So, when you come to the banks of the raging Jordan rivers in your life, you will not be intimidated. You will not be afraid to choose the Word of God.

If you don't know what the Word of God says, you won't know what the Word of God can do. And if we don't know what the Word of God can do, we will lean to our own understanding, where the Word of God says lean not to your own understanding.

When we acknowledge in our thought process what society has taught us, where the Word of God says to "acknowledge Him in all thy ways," our direction will be down a wrong path.

The Word says He shall direct our path. It is time to cross "The Jordan Rivers" in our lives.

Understand that when we look back upon our life, reality tells us we are where we are because of the choices that we have chosen. But that is nothing new. God gave man the ability to choose, and the reality is, we are who we are because of our choices.

The choices we choose when we believe the truth that God has already made a way for, we will react and respond and receive the benefits from those actions.

John 1:1 where the Word was made flesh, instructions and information that when followed will lead us from the captivity of what society has taught and shown us; and lead us from our sinful nature we were born into.

But if we choose to ignore the information, we will continue to trip over our fleshly desires.

As we close, understand we are put into a position to make choices between right and wrong, and good and evil; to trip in our flesh or walk in the Spirit of God.

We are put in a position to make choices and show if we are worthy of the promises of God and put in a position to show equal value in our praying...

A position to show equal value in our praising...

An equal value of faithfulness, and an equal value in our fasting and tithing.

Somebody here today is in a position to show if you are worthy of the promises of God. Abraham was put in a position to choose and show it with his son Isaac.

Job was put in a position to choose and show.

Adam and Eve were put in a position to choose and show.

Peter was put in a position to choose when the Lord asked him three times, "Peter do you love me more than these?"

And Peter answered, "You know I love you, Lord."

And that's how you cross over to the other side of the Jordan Rivers in today's reality-- by loving God more than these; more than these things of the world.

By loving God more than the things that are seen in the world, and what society has shown you; By understanding that things of this world are temporary, and the things of God are everlasting.

When you stop tripping in your flesh ... your anger cannot last.

When you stop tripping in your flesh ...your hatred cannot last.

When you stop tripping in your flesh ... your jealousy cannot last.

When you stop tripping in your flesh...you won't even be able to gossip as much...and all the honest people said, "Amen."

Help us, Holy Spirit--- I am not going to talk about my brother or sister while they are tripping in their flesh. I am going to pray that Your Grace and Mercy be sufficient until they are able to walk in Your Spirit.

I am going to pray, knowing that, in that prayer, I will create a Spiritual connection of Oneness with the Spiritual nature that dwells in that particular human body.

What You did for me, Lord, while I was in my mess. I send that same energy of Love their way; I direct it their way, and be a comfort, be a help, be a healing source to whatever negative forces are out there in that situation.

You will do for them, Lord what You did for me, even though they may have done me wrong,

I love You, Lord, and I am not going to trip in my own flesh, but I am going to walk in Your Loving Spirit. Because it is worth it to see You in Glory.

Because it worth it to hear You say, "Well done... My good and faithful servant." And the honest people said? "Amen."

This text is about crossing the Jordan River. A wide range of space from this side to the other side.

I believe the widest space we have to cross in today's reality is the wide space of doubt and confusion in our mentality and thought process.

We have a wide space of doubt and confusion from the natural to the spiritual.

A wide space from looking at life from society's point of view to looking at life from God's point of view.

It seems that some are hesitant, or have an unwillingness, to crossover from living for the world, to living for Jesus.

And that space of doubt and confusion has a direct effect on man's free will and will influence what he sees.

It will influence what he feels, and what he believes; what he chooses, and ultimately his place in the everlasting.

That's how some people see what they want to see and believe what they want to believe.

As for me and my free will, I might have been turned around as a child and taught east was west, and west was east, but I thank God He shined His Light on me.

And I was able to learn to accept the real truth.

The truth is that east was east, and west was west.

I was able to accept that the truth is, it is better to please God than to please man.

I was able to accept the truth, and believe, that God can do all everything, and I can do all things through Christ who strengthens me.

So, when I come to the banks of the Jordan River, the width and the space of doubt and confusion in my thought process is bridged by my belief that Jesus died on the cross,
  went down below,

got the keys,
came back up,
and proclaimed: ALL power is in My Hands."

Now I've got the power to cross the Jordan rivers in my life.

Has anybody else in here got the power to cross over to the other side?!

Can I get some honest people to stand up with me?

Can I get some honest people to cross over to the other side with me?

I came to shed light on what you already know, and that is, that the adversary does not want you crossings over to the other side, to the land of milk and honey, to experience life in Christ Jesus. Understand, the same God that told Joshua, "I will have the same relationship with you that I had with Moses," would have the same relationship with *US*.

If you just follow Jesus, you will have good success in crossing the Jordan Rivers.

Time to stop tripping in your flesh.

Time to start understanding what you did not understand before.

Thoughts lead to actions, actions lead to habits, and habits lead to character, ultimately leading to destiny.

God's brain, God's mind, God's body.

Thoughts like Jesus lead to actions like Jesus. Actions like Jesus lead to habits like Jesus, and habits like Jesus' lead to character like Jesus. Character like Jesus leads to the same destiny as Jesus.

"So I say, walk by the Spirit, and you will not gratify the lusts of the flesh." *(Galatians. 5.16)*

1st Corinthians tells us that the Spiritual Man searches all things, *The deep things* of God, and glory to God! This is a congregation diligently seeking …the deep things of God

And all the honest people said Amen.

This is the deepness that this generation of Godly people must understand.

What is it worth, in response to the issues of life?

Is it worth saying something back?

Is it worth retaliating?

What is the value in it for me in the long run?

If you can see & understand the problem (the Jordan River), you should also be able to see and understand the reward of overcoming the problem.

Crossing over that Jordan River. Don't be Afraid to Trust God.

The Epilogue

The things God has spoken into the mind of humanity, are in the mind of humanity today, residing on what we call a subconscious level.

They are willing and waiting to be activated by words, labels, and phrases that will make the human mind visualize and see the positive increase that will be placed in the forethought of the human's mind.

Free will to choose and manifest this forethought-- the law of attraction, visualization principles, the path of least resistance-- are still running the way God the Creator of all Heaven and Earth designed them to run in, God's Brain, God's Mind, and God's Body that He designed, it is Man who gets to choose what he puts in it.

These *Lessons From The School of Hard Knocks* are the things that have been spoken while being present with yourself.

They are subconscious behaviors being brought up for a conscious conversation; the relationship between the human mind and God, the Source Who is not visible with natural eyes.

How to walk in your spirit and stop tripping in your flesh.

"*I Am, That I Am.*"

On the count of three, I want you to close your eyes and say, "I Am, That I Am."

I want you to say it without moving your mouth --- 1, 2, 3:

"I Am, That I Am."

In your consciousness of being wide awake and alert, what was that voice that you heard in your consciousness?

What voice just spoke when you repeated those words without any sound coming out of your mouth; without any vibrations going into your ears for you to hear?

Your mouth did not speak, and your ears did not hear, but you heard yourself say, "I am that I am."

If you were to close your eyes again, you can still see an image that says, "I Am that I Am"

What is the sight that you are seeing with and getting information from?

What is the sight that can see beyond time?

What is the sight that can look back into time?

The Epilogue

You can see your past, but you don't use these two eyes to do so.

You can hear the past, but you don't use your two ears to hear it.

The Breath of God states it clear as day

I Am That I Am

I am *Not* the flesh; the flesh is the dirt of Earth.

I am the Energy source that has the consciousness of being in the flesh of this earth suit.

It is not a big stretch to visualize the flesh as a suit or an outer protectant layer encasing energy.

What is the sight that can see your future?

What is the sight that can see beyond time?

The Breath that God the Source, breathed into the nostrils of Adam.

This is an actual fact. You can speak without moving your mouth, and you can hear without ears.

So, what happens when these physical ears, eyes, mouth, go into the soil of earth from which it came, When these other senses go into the grave, returning to dust, all you're left with is the breath of God. The energy source that is more powerful than anything we could imagine.

The "I," the Inner voice inside, the energy source of the visions of the past, and visions of the future.

We can close our eyes right now and have a conversation with "me, myself, and I," and ask them to come to a conscious agreement: "As I Am in God, GOD is in me."

The Fifth Dimension
The I that is I, is the I that is We

We are taught in four dimensions: front-back, left-right,

up-down, and time.

- We were taught how to read from left to right.
- We were taught how to look up and down
- We were taught how to go to the front or go to the back.
- We were taught know how to tell time

But now ask your yourself, were you taught how to go in and out of the fifth dimension of reality, how to go in and out of things understanding how to move in and out of your spiritual awareness were you taught how to pass through, just as the Man Christ Jesus was able to pass through with his Spiritual awareness of The Power of God the Creator of All Heaven and Earth within.

Let us develop some neural pathways that go in and out, and neural pathways that understand that I exist without the body: neural pathways that have a better understanding that the body without the spirit is just a pile of dirt.

The Epilogue

Let us experience mental stimulation for positive increase and at the same time mental stimulation on how to prevent taking an *L*, or a loss.

Let us develop the ability to grow neural pathways of 'in and out' for the Spirit of Man that is on the inside, to be manifested in these bodies, on the outside.

How the Spirit of Man moves through these bodies, taking in things from the outside world and filtering them through God's brain, and for God's mind, to be manifested in God's body.

The in and out of things not seen with the natural eye.

The in and out of things not vibrated or heard with the natural ear.

The in and out of things not said with your natural lips moving.

We are taught up and down, left to right, and front and back, and to tell time; but not in and out.

The curriculum is about building and applying continuous improvement methods for neural pathways that create the in and out the going through as Jesus went through the midst of them, and "so passed by, passing through their midst, He went away."

Knowing how to move in and out of things is possessing neural pathways that allow the source on the inside to control the body on the outside.

How did Jesus do it? By proclaiming, "I AM, THAT I AM." This is how the Jesus (the man portion did it) so, I started doing it too.

As I look back from where He has brought me from, I can recall how I came out of darkness... Into His marvelous light,

It was the "In and Out of my relationship with the internal Power of God, from God, The I AM THAT I AM,"

I remember the many different times when things should have happened a lot worse and yet didn't happen.

To the point, using the logic of Zeros and Ones that pointed towards God, I was able to shut down the neural pathways that were in Love with "free-basing cocaine" and, I created some new ones that that said "Love God more, because that's the dopamine drop, which comes from the brain that God designed to release dopamine in the first place." #There's no high like The Most High!

Gods Brain for Gods Mind is designed to create mathematical equations that are turned into human movements of what the hands do, where your feet go, and what your mouth says with regards to manifesting the likeness and image of Love for the outside world to see,

When society says there is no way out LFTSOHKS is the handbook of how Gods Brain for Gods Mind is designed to create a way ...out of no way.

The Epilogue

The chaos we see today has shown enough evidence, for the statement to be said, that "This is a crazy world we live in!"

LFTSOHKS is designed to reverse-engineer the neural network that has been taught and replicated by repetition of the mathematical equations to produce the negative production, in the human thought that comes from the thought process of fallen Adam.

LFTSOHKS is a book full of words, labels, and phrases designed for to make the human mind visualize what's best for your future self in "this crazy world we live in."

Lessons From the School of Hard Knocks

LFTSOHKS teaches the in and out of things creating the creativity in the mind to manifest the "way out of no way" that the confusion of societal norms through mental manipulation of words, labels, and phrases, designed to make the human mind visualize, what is best for his future self.

I was able to shut down the neural pathways that were in Love with "free-basing cocaine" and, I created some new ones that that said "Love God more, because that's the dopamine drop, which comes from the brain that God designed to release dopamine in the first place." This is how God can take the taste right out of your mouth, without setting a foot in a Betty Ford Clinic.

To the Point. There is Power in The Word, Power in A Word, Power in the Words, Labels, and Phrases that we tell ourselves that will control what the body does.

How the taste was taken out of my mouth: by remembering how bad the taste would taste, before I tasted it, the consciousness of how to stop it before it starts, the foresight of seeing into your future by looking at your past is an example of the In and Out, by remembering how bad the taste tasted before I tasted it.

Remembering the feeling and feeling it... before I had to feel I t

Remembering the conclusion of how I felt before the conclusional effects

Remembering how much it broke my finances, before I had to be broke

None of these things, was I able to do with my natural senses:

I received a visual image that I did not see, with my natural eyes

Acts of Kindness we call them, but they are examples of the IN and Out of the Energy Source we call Love

In and Out is the ability to stay focused on creating positive energy, which can be created by multiplying positive energy that overcomes the bad habit that was taught, whether it was taught you by someone else or it was taught to you by your own self.

These are the neural pathways that should have been created and taught during formative years

The In and Out Power of the Energy Force we Call Love

The Epilogue

It was after my recovery phase. I had an encounter with a friend of mine who had fallen out of touch with reality and began wearing kung fu outfits to those of "ShoNuff" in the movie *The Last Dragon*. He was way out of control mentally, taking drugs and not paying back debts, but still doing drugs, but because he had the appearance that he knew some martial arts and they did not mess with him too much.

But he was a good friend and so, after I had cleaned myself up (Thank You Lord), I was checking on my brother because we had had some good times getting high and I considered him a friend, and this what friends are supposed do.

It so happened that the day that we did meet up and I remember it very well, we met up and he had on his Karate outfit and had on the pointed straw woven hat, that that covers a part of your face.

At the dusk of the Sun, we were walking down Fulton Street by the Franklin Avenue shuttle. We were crossing the street when about 15 "bad guys," several who he had owed money to, jumped him and his fake karate did not work and just one of them was going to hit in in the head with a bat. And just as one of them swung for contact I stopped the bat from striking my friend's head, then I thought "OMG followed by a WTH did I just do. I am getting ready to get the 'Wrath of Kahn' stomped on my donkey part" because basic street code dictates if you are not for us you are against us, so in my conscious mind I was supposed to get it too. When I stopped the bat everyone else stopped their hitting, as well, and I remember the looks on their faces that said, with full respect, "Did this N-Word," #"NEGUS "really just do what we think he did." {stop them from killing this man, because that's what they thought they were supposed to do, to get to keep their "society taught "street credibility} ...and they just left us both alone in the middle of Fulton St.

Word Check: NEGUS

definition of Negus: Royalty, Kings, and Emperors of Ethiopia

#2015 Oprah Winfrey, Jay-Z, Kendrick Lamar, and the word "Nigger v Negus" #taught to recall of negative emotion from the taught meaning on purpose.

#The In and Out of the Power of God produces Acts of Kindness

This experience changed the neural circuitry in my mind.

The premise of this statement and the logic of "I AM THAT I AM" in which I claim myself to be, literally and physically shut down a whole lot of neural pathways that were not beneficial for my future self, because they did not align up with the "I AM THAT I AM," of who I claimed myself to be.

Affirmations are the stories we tell ourselves with repetition for replication of the story we tell ourselves to be manifested in the bodies that God made knowing we are strengthening the neural network for His Glory and Purpose.

"Greater is He that is in me, than he that is in the world".

The Epilogue

"I am the best person in the world at being me."

"I AM THAT I AM" and that you will be, as you are in God, God is in you.

#OneBreath #One Love

Once you discover that level of consciousness, where you identify with the Spirit of Man inside the body, it gives Power to the Body. Now when you speak, it is no longer the outside body speaking but the inside Spirit of Man.

The Spirit of Man speaks, not the flesh, It is not just a body speaking, but the Spirit of Man speaking through the Body.

Now you can command the Energy that is within and throughout the universe because it is no longer just the flesh on man speaking, but the Breath of God speaking.

It is the spiritual and measured essence that is God, the Creator of all Heaven and Earth.

The Current, the Power, and the Moving Force of God going In and Out of all mankind.

The Breath/Pneuma Spirit God breathed into mankind and made mankind a living soul.

That same measure of Power is in every omnipresent breath that we breathe in today.

Every breath we take today is for the same purpose --- to make these bodies a living soul, with abilities to create and heal the body mentally from Adam's mental decision.

LFTSOHKS offers the tools necessary to create new neural pathways; ways to heal God's Brain, for God's Mind, for God's Body.

The Power that comes from the Love of God, the Source, is transferred through our free will to choose.

It is my conclusion that the path of least resistance for the Spirit of Man to master the human experience is to use my God-given gift of free will to choose who I AM and be my most authentic self.

Using the words, labels, and phrases of *The Basic Instructions Before Leaving Earth*, helps me see more clearly how to filter my human emotions, through the Thalamus Reticular Activating System Region for the correct variables and equations of input that will produce positive production in my manifested actions to go forth with, to multiply with to replenish with and to subdue with all the things that I take in from the outside world and Manifest into my Action, The Energy Force of my creation.

Consciousness of the Love of God is having the foresight of things that don't line up with the Love of God.

It is the ability to know the end from the beginning, and how to now use your free will to choose more clearly and more often, along with understanding how to *act* like you know the end from the beginning.

The wonderful thing about this book is, "...if you didn't know, now you know!"

For the flesh lusts against the Spirit, and the Spirit against the flesh; and the two don't like each other. If your flesh is in a rush to do something so fast, react so quickly, it should be a yellow flag, because we already know the flesh of man is inclined to work against the Spirit of Man.

Knowing what *time* it is with our own will, and listening when wisdom speaks and signals a yellow, or red flag, proceed with caution. Let this cup, pass from me, the pain, and the confusion.

The tricks and the traps of the adversary, and the issues of life,

If it can be possible. If it can be possible? It is possible in your belief system. It is possible to manifest in the members of your body..

There is power in the word of God. *It* is where we make an insertion, in reference to the story we tell ourselves. By removing the limitations of our will for righteousness and positivity to know God, we connect to the unlimited love, and the power that lies within.

There is power in the Word of God. And it is when we insert it in reference to the story we tell ourselves that it removes limitations from our free will for righteousness and positivity to know God. We will get a stimulation and visual image of how to connect to the unlimited power that connected to the Word Love, and the Power that lies within.

"These things I have spoken to you while being present with you. But the Helper, the Holy Spirit, whom the Father will send in My name, He will teach you all things, and bring to your remembrance all things that I said to you."

-The End or Better yet The Beginning

Bibliography

Biblical Citizenship in Modern America (c) 2021 by Rick Green

Coots, J. Fred, and Haven Gillespie. *Santa Claus Is Comin' to Town*. Oct. 1934.

"Definition of FORESIGHT." *Www.merriam-Webster.com*, www.merriam-webster.com/dictionary/foresight. Merriam-Webster Dictionary.

Frantz Fanon, et al. *Alienation and Freedom*. 1960. London, Bloomsbury Academic, 2008.

Green, Al. *Love and Happiness*. 1972.

Green, Rick, Biblical Citizenship in Modern America, 2021 https://www.rccgvictorycentre.com. "*Attitude & the Love of God* - Pastor Akinola,." RCCG Victory Centre, 12 July 2012, www.rccgvictorycentre.com/attitude-the-love-of-god/. Accessed 6 Feb. 2022.

The Last Dragon, Tri- Star Pictures Barry Gordy 1985

Snap. *The Power*. 1989. "World Power" Studio Album

Lamar, Kendrick. *The Blacker the Berry*. 2015. "To Pimp A Butterfly" Studio Album.

Lox, The. *Money, Power, & Respect*. Mar. 1998.

"New King James Version (NKJV) - Version Information - BibleGateway.com." *Www.biblegateway.com*, www.biblegateway.com/versions/New-King-James-Version-NKJV-Bible/. Gal. 5.16-17, Pro. 3.5-6, Pro. 4.7, Pro. 23.7, Phi. 4.13, Gen. 1.2, Mat. 11.28, 1 Joh. 4.4, 1 Cor. 12.12, Rom. 7.15-25, Joh. 14.26.

"Newton's Third Law of Motion." *Nasa.gov*, 2014, www.grc.nasa.gov/WWW/K-12/rocket/newton3r.html.

Potter, Jessie. -. 7th Annal Woman to Woman Conference. Milwaukee, Wisconsin -1981.

Sartre, Jean-Paul. *Dirty Hands*. 1946. Act 5, Scene 3.

Shakur, Tupac. *Thug Life*. 1994.

Team soul. "Swami Satchidananda." *Fearless Soul - Inspirational Music & Life Changing Thoughts*, 17 July 2018, iamfearlesssoul.com/power-of-the-mind-quotes/. Quote.

"U.S. Constitution - Thirteenth Amendment | Resources | Constitution Annotated | Congress.gov | Library of Congress." *Constitution.congress.gov*, constitution.congress.gov/constitution/amendment-13/.

Wilkerson, Isabel. *Caste: The Origins of Our Discontents*. New York, Random House, 4 Aug. 2020, pp. 3–20.

X, Malcolm. *By Any Means Necessary*. Organization of Afro-American Unity Founding Rally. Washington Heights, Manhattan, NY.